LIFE ON A
LITTLE-KNOWN
PLANET

LIFE ON A LITTLE-KNOWN PLANET

DISPATCHES FROM A CHANGING WORLD

ELIZABETH KOLBERT

CROWN
NEW YORK

CROWN
An imprint of the Crown Publishing Group
A division of Penguin Random House LLC
1745 Broadway
New York, NY 10019
crownpublishing.com
penguinrandomhouse.com

All of the essays in this work were originally published in *The New Yorker* except for "The Guru of Doo-Doo: Profile of Sam Wasser, Who Uses DNA to Fight Elephant Poaching," which was originally published as "The Elephant Detective" in the January–February 2017 issue of *Smithsonian Magazine.*

Library of Congress Cataloging-in-Publication Data has been applied for.

Hardcover ISBN 979-8-217-08606-1
Ebook ISBN 979-8-217-08607-8

Editor: Gillian Blake
Editorial assistant: Jessica Jean Scott
Production editor: Terry Deal
Text designer: Aubrey Khan
Production: Christopher Andrus
Copy editor: Elisabeth Magnus
Proofreaders: Jill Twist and Chuck Thompson
Publicist: Gwyneth Stansfield
Marketer: Chantelle Walker

Space stars background on pages ii, 1, 79, 157, and 227: Shutterstock / KK.KICKIN

Manufactured in the United States of America

1st Printing

First Edition

The authorized representative in the EU for product safety and compliance is Penguin Random House Ireland, Morrison Chambers, 32 Nassau Street, Dublin D02 YH68, Ireland, https://eu
-contact.penguin.ie.

CONTENTS

PART ONE
Creatures Great and Small

•

PART TWO
A Sense of Place

•

PART THREE
Big Ideas

•

PART FOUR
All We Can Save

•

LIFE ON A
LITTLE-KNOWN
PLANET

INTRODUCTION

A FEW MONTHS BEFORE the world shut down for COVID, I flew from the city of Adelaide, on Australia's southern coast, to the town of Roxby Downs, in the country's dusty interior. Nearly everyone who lives in Roxby Downs, or even just visits it, is connected to a copper and uranium mine north of town. I had come not for the metals, though, but for the marsupials. That night, I slept in a trailer in the outback. I had the place to myself except for a quoll that was occupying the living room. Quolls are cute but ferocious creatures that look a bit like spotted ferrets. They're nocturnal, so all night long my neighbor paced her pen. Several times when I heard her scrabbling around, I got out of bed and went to check on her. I felt sad for her yet also exhilarated. Quolls are so rare these days that most Australians will see them only in zoos. Here I was, just passing through and practically sharing a flat with one.

The following day, the quoll was released by an ecologist named Katherine Moseby. For the past three decades, Moseby has been trying to protect Australia's marsupials by putting up fences. The resulting exclosures, as the fenced-in spaces are known, might be thought of as zoos turned inside out: while zoos keep rare animals in, exclosures keep common animals out. Moseby's fences are specially designed to exclude cats, which were introduced to Australia in the early nineteenth century and now roam the countryside by the millions. Predation by feral cats is one of the main reasons Australia's fauna is in so much trouble: of the country's two hundred and seventy endemic species of land mammals, more than thirty are now extinct. Another fifty, including eastern quolls and northern quolls, are endangered.

By the time I got to Roxby Downs, Moseby had managed to fence in—or really, I suppose, out—an area of almost fifty square miles. Additional fencing divided this area into six "paddocks," some of which could be used for conservation and some for experimentation. One experiment—the one that had drawn me to the outback—involved greater bilbies, rabbit-size marsupials with shrew-like noses and rat-like tails. Moseby thought that by exposing them to just a few cats at a time, it might be possible to nudge evolution along and, over generations, produce a line of bilbies with an innate fear of felines.

"We're not going to ever get rid of every cat in the whole of Australia," she observed at one point. At another, she said: "People say to me, 'Oh, this could take a hundred years.' And I say, 'Yeah, it could take a hundred years. What else are you doing?'"

WE LIVE IN an extraordinary time. This is true not just from a parochial American (or Australian) perspective but from a planetary one. A recent study suggests that all living things—you, me, the birds and the bees, the snake in the grass, the worm in the apple as well as the tree the apple didn't fall far from—are descended from a single-celled organism that evolved around four billion years ago. Since the days of LUCA—short for the Last Universal Common Ancestor—a great deal has happened. The atmosphere has been suffused with oxygen. The continents have crashed together and drifted apart. The entire planet has frozen over and thawed out. Plants and animals have crept out of the water and onto land. The age of reptiles has given way to the age of mammals.

But over the last four billion years, only very rarely has change rushed along at the pace it is moving today. And probably never before have so many different forms of change been operative. Owing to the actions of one species, which is, of course, our own, the world is now heating up at the same time that the oceans are acidifying, forests are shrinking, groundwater is disappearing, microplastics are proliferating, and so-called forever chemicals are spreading. Every day, people are transporting thousands of species around the world, mostly in cargo. Those that find their new homes congenial—think cats in Australia—will often go on to wreak havoc. Many geologists believe that humans have become so much the dominant force on the planet that we have entered a new epoch: the Anthropocene.

Meanwhile, even as we are upending the natural world, we are coming to understand it in ever more intimate detail. In the Anthropocene, scientists can tease DNA from the bones of animals, such as mammoths, that vanished thousands, or even tens of thousands, of years ago. They can track bird migrations and butterfly movements by means of tiny transponders. They can piece together the history of the climate from shells found at the bottom of the sea and the history of the seas from the magnetic orientation of the oceanic crust.

E. O. Wilson once described the "human trajectory"—increasingly destructive in its impacts, increasingly sophisticated in its insights—as the "ultimate irony of organic evolution." It is this irony that animates this book. Almost all of the stories collected here originally appeared in *The New Yorker*, where I've been fortunate enough to work for the last twenty-five years. These have been bad years for much of life on earth and all-absorbing ones for people who write about life on earth. Reporting on the "human trajectory" has taken me to the Greenland ice sheet, the Peruvian Andes, and the South Island of New Zealand, among many other spectacular places. In the course of my reporting, I have watched entirely new species be discovered and have visited with the last living members of species that once were widespread. I have also met scores of brilliant and dedicated people who are working to get the world on a better trajectory; these include researchers like Moseby, as well as activists, amateur naturalists, and ordinary citizens. Many appear in the pages that follow. Many others have informed my thinking but remain unnamed. One of the great privileges of being a journalist is it gives you the license to phone pretty much anyone, pretty much anywhere. I am very grateful to all those who answered my calls.

• • •

FINALLY, A WORD about words. In the grand scheme of things, words are latecomers. If you imagine the history of the planet compressed into a single day that began at twelve a.m., then LUCA showed up surprisingly early on, at about two in the morning. The first fish didn't swim onto the scene until nighttime, and the first mammals didn't appear until almost eleven p.m. No one knows whether early humans like *Homo erectus* could talk, but if so, language originated about half a minute ago. If not, then words have, figuratively speaking, been around for only about ten seconds. Written words have been around for maybe a tenth of a second.

Trying to capture in words a world that for billions of years did just fine without them is, admittedly, a tricky enterprise. "You cannot get the desert into a book any more than a fisherman can haul up the sea with his nets," Edward Abbey observed in a book about the desert. The same holds true for a Caribbean bay or an Alpine glacier or a patch of scrubland in West Texas. Humans perceive only a sliver of the globe's (nearly) infinite variety, and only a sliver of that sliver can be condensed into sentences. At the same time, narrative is what people use to make sense of things. It is how we convey to each other what we think is funny or important or urgent or tragic. Since the natural world cannot tell its own stories, it falls to us to step in. The essays and articles that follow represent my effort to do so.

PART ONE

CREATURES GREAT
AND SMALL

TALK TO ME

Can Artificial Intelligence Allow Us
to Speak to Other Species?

DAVID GRUBER BEGAN his almost impossibly varied career studying bluestriped grunt fish off the coast of Belize. He was an undergraduate, and his job was to track the fish at night. He navigated by the stars and slept in a tent on the beach. "It was a dream," he recalled. "I didn't know what I was doing, but I was performing what I thought a marine biologist would do."

Gruber went on to work in Guyana, mapping forest plots, and in Florida, calculating how much water it would take to restore the Everglades. He wrote a PhD thesis on carbon cycling in the oceans and became a professor of biology at the City University of New York. Along the way, he got interested in green fluorescent proteins, which are naturally synthesized by jellyfish but, with a little gene editing, can be produced by almost any living thing, including humans. While working in the Solomon Islands, northeast of Australia, Gruber discovered dozens of species of fluorescent fish, including a fluorescent shark, which opened up new questions. What would a fluorescent shark look like to another fluorescent shark? Gruber enlisted researchers in optics to help him construct a special "shark's eye" camera. (Sharks see only in blue and green; fluorescence, it turns out, shows up to them as greater contrast.) Meanwhile, he was also studying creatures known as comb jellies at the Mystic Aquarium, in Connecticut, trying to determine how, exactly, they manufacture the molecules that make them glow. This led him to wonder about the way that jellyfish experience the world. Gruber enlisted another set of collaborators to develop robots that could handle jellyfish with jellyfish-like delicacy.

"I wanted to know: Is there a way where robots and people can be brought together that builds empathy?" he told me.

In 2017, Gruber received a fellowship to spend a year at the Radcliffe Institute for Advanced Study, in Cambridge, Massachusetts. While there, he came across a book by a free diver who had taken a plunge with some sperm whales. This piqued Gruber's curiosity, so he started reading up on the animals.

The world's largest predators, sperm whales spend most of their lives hunting. To find their prey—generally squid—in the darkness of the depths, they rely on echolocation. By means of a specialized organ in their heads, they generate streams of clicks that bounce off any solid (or semi-solid) object. Sperm whales also produce quick bursts of clicks, known as codas, which they exchange with one another. The exchanges seem to have the structure of conversation.

One day, Gruber was sitting in his office at the Radcliffe Institute, listening to a tape of sperm whales chatting, when another fellow at the institute, Shafi Goldwasser, happened by. Goldwasser, a Turing Award–winning computer scientist, was intrigued. At the time, she was organizing a seminar on machine learning, which was advancing in ways that would eventually lead to ChatGPT. Perhaps, Goldwasser mused, machine learning could be used to discover the meaning of the whales' exchanges.

"It was not exactly a joke, but almost like a pipe dream," Goldwasser recollected. "But David really got into it."

Gruber and Goldwasser took the idea of decoding the codas to a third Radcliffe fellow, Michael Bronstein. Bronstein, also a computer scientist, is now the DeepMind Professor of AI at Oxford.

"This sounded like probably the most crazy project that I had ever heard about," Bronstein told me. "But David has this kind of power, this ability to convince and drag people along. I thought that it would be nice to try."

Gruber kept pushing the idea. Among the experts who found it loopy and, at the same time, irresistible were Robert Wood, a roboticist at Harvard, and Daniela Rus, who runs MIT's Computer Science and Artificial Intelligence Laboratory. Thus was born the Cetacean Translation Initiative—Project CETI for short. (The acronym is pronounced "setty," and purposefully recalls SETI, the Search for Extraterrestrial Intelligence.) CETI represents the most ambitious, the most technologically sophisticated, and the most well-funded effort ever made to communicate with another species.

"I think it's something that people get really excited about: Can we go from science fiction to science?" Rus told me. "I mean, can we talk to whales?"

SPERM WHALES ARE NOMADS. It is estimated that, in the course of a year, an individual whale swims at least twenty thousand miles. But scattered around the tropics, for reasons that are probably squid-related, there are a few places the whales tend to favor. One of these is a stretch of water off Dominica, a volcanic island in the Lesser Antilles.

CETI has its unofficial headquarters in a rental house above Roseau, the island's capital. The group's plan is to turn Dominica's west coast into a giant whale-recording studio. This involves installing a network of underwater microphones to capture the codas of passing whales. It also involves planting recording devices on the whales themselves—cetacean bugs, as it were. The data thus collected can then be used to "train" machine-learning algorithms.

In July, I went down to Dominica to watch the CETI team go sperm whale bugging. My first morning on the island, I met up with Gruber just outside Roseau, on a dive-shop dock. Gruber, who is fifty, is a slight man with dark curly hair and a cheerfully anxious manner. He was carrying a waterproof case and wearing a CETI T-shirt. Soon, several more members of the team showed up, also carrying waterproof cases and wearing CETI T-shirts. We climbed aboard an oversize Zodiac called *CETI 2* and set off.

The night before, a tropical storm had raked the region with gusty winds and heavy rain, and Dominica's volcanic peaks were still wreathed in clouds. The sea was a series of white-fringed swells. *CETI 2* sped along, thumping up and down, up and down. Occasionally, flying fish zipped by; these remained aloft for such a long time that I was convinced for a while they were birds.

About two miles offshore, the captain, Kevin George, killed the engines. A graduate student named Yaly Mevorach put on a set of headphones and lowered an underwater mike—a hydrophone—into the waves. She listened for a bit and then, smiling, handed the headphones to me.

The most famous whale calls are the long, melancholy "songs" issued by humpbacks. Sperm whale codas are neither mournful nor musical. Some people compare them to the sound of bacon frying, others to popcorn popping. That morning, as I listened through the headphones, I thought of horses clomping over cobbled streets. Then I changed my mind. The clatter was more mechanical, as if somewhere deep beneath the waves someone was pecking out a memo on a manual typewriter.

Mevorach unplugged the headphones from the mike, then plugged them into a contraption that looked like a car speaker riding a broom handle. The contraption, which I later learned had been jury-rigged out of, among other elements, a metal salad bowl, was designed to locate clicking whales. After twisting it around in the water for a while, Mevorach decided that the clicks were coming from the southwest. We thumped in that direction, and soon George called out, "Blow!"

A few hundred yards in front of us was a gray ridge that looked like a misshapen log. (When whales are resting at the surface, only a fraction of their enormous bulk is visible.) The whale blew again, and a geyser-like spray erupted from the ridge's left side.

As we were closing in, the whale blew yet again; then it raised its elegantly curved flukes into the air and dove. It was unlikely to resurface, I was told, for nearly an hour.

We thumped off in search of its kin. The farther south we traveled, the higher the swells. At one point, I felt my stomach lurch and went to the side of the boat to heave. "I like to just throw up and get back to work," Mevorach told me.

• • •

TRYING TO ATTACH a recording device to a sperm whale is a bit like trying to joust while racing on a Jet Ski. The exercise entails using a thirty-foot pole to stick the device onto the animal's back, which in turn entails getting within thirty feet of a creature the size of a school bus. That day, several more whales were spotted. But, for all of our thumping around, *CETI 2* never got close enough to one to unhitch the tagging pole.

The next day, the sea was calmer. Once again, we spotted whales, and several times the boat's designated pole handler, Odel Harve, attempted to tag one. All his efforts went for naught. Either the whale dove at the last minute or the recording device slipped off the whale's back and had to be fished out of the water. (The device, which was about a foot long and shaped like a surfboard, was supposed to adhere via suction cups.) With each new sighting, the mood on *CETI 2* lifted; with each new failure, it sank.

On my third day in Dominica, I joined a slightly different subset of the team on a different boat to try out a new approach. Instead of a long pole, this boat—a forty-foot catamaran called *CETI 1*—was carrying an experimental drone. The drone had been specially designed at Harvard and was fitted out with a video camera and a plastic claw. Because sperm whales are always on the move, there's no guarantee of finding any; weeks can go by without a single sighting off Dominica. Once again, though, we got lucky, and a whale was soon spotted. Stefano Pagani, an undergraduate who had been brought along for his piloting skills, pulled on what looked like a VR headset, which was linked to the drone's video camera. In this way, he could look down at the whale from the drone's perspective and, it was hoped, plant a recording device, which had been loaded into the claw, on the whale's back.

The drone took off and zipped toward the whale. It hovered for a few seconds, then dropped vertiginously. For the suction cups to adhere, the drone had to strike the whale at just the right angle, with just the right amount of force. Post impact, Pagani piloted the craft back to the boat with trembling hands. "The nerves get to you," he said.

"No pressure," Gruber joked. "It's not like there's a *New Yorker* reporter watching or anything." Someone asked for a round of applause. A cheer went up from the boat. The whale, for its part, seemed oblivious. It lolled around with the recording device, which was painted bright orange, stuck to its dark-gray skin. Then it dove.

Sperm whales are among the world's deepest divers. They routinely descend two thousand feet and sometimes more than a mile. (The deepest a human has ever gone with scuba gear is just shy of eleven hundred feet.) If the device stayed on, it would record any sounds the whale made on its travels. It would also log the whale's route, its heartbeat, and its orientation in the water. The suction was supposed to last around eight hours; after that—assuming all went according to plan—the device would come loose, bob to the surface, and transmit a radio signal that would allow it to be retrieved.

I said it was too bad we couldn't yet understand what the whales were saying, because perhaps this one, before she dove, had clicked out where she was headed.

"Come back in two years," Gruber said.

EVERY SPERM WHALE'S tail is unique. On some, the flukes are divided by a deep notch. On others, they meet almost in a straight line. Some flukes end in points; some are more rounded. Many are missing distinctive chunks, owing, presumably, to orca attacks. To ID a whale in the field, researchers usually rely on a photographic database called Flukebook. One of the very few scientists who can do it simply by sight is CETI's lead field biologist, Shane Gero.

Gero, who is forty-three, is tall and broad, with an eager smile and a pronounced Canadian accent. A scientist-in-residence at Ottawa's Carleton University, he has been studying the whales off Dominica since 2005. By now, he knows them so well that he can relate their triumphs and travails, as well as who gave birth to whom and when. A decade ago, as Gero started having children of his own, he began referring to his "human family" and his "whale family." (His human family lives in Ontario.) Another marine biologist once described

Gero as sounding "like Captain Ahab after twenty years of psycho-therapy."

When Gruber approached Gero about joining Project CETI, he was, initially, suspicious. "I get a lot of emails like 'Hey, I think whales have crystals in their heads' and 'Maybe we can use them to cure malaria,'" Gero told me. "The first email David sent me was, like, 'Hi, I think we could find some funding to translate whale.' And I was, like, 'Oh, boy.'"

A few months later, the two men met in person, in Washington, D.C., and hit it off. Two years after that, Gruber did find some funding. CETI received thirty-three million dollars from the Audacious Project, a philanthropic collaborative whose backers include Richard Branson and Ray Dalio. (The grant, which was divided into five annual payments, will run out in 2025.)

The whole time I was in Dominica, Gero was there as well, supervising graduate students and helping with the tagging effort. From him, I learned that the first whale I had seen was named Rita and that the whales that had subsequently been spotted included Raucous, Roger, and Rita's daughter, Rema. All belonged to a group called Unit R, which Gero characterized as "tightly and actively social." Apparently, Unit R is also warmhearted. Several years ago, when a group called Unit S got whittled down to just two members—Sally and TBB—the Rs adopted them.

Sperm whales have the biggest brains on the planet—six times the size of humans'. Their social lives are rich, complicated, and, some would say, ideal. The adult members of a unit, which may consist of anywhere from a few to a few dozen individuals, are all female. Male offspring are permitted to travel with the group until they're around fifteen years old; then, as Gero put it, they are "socially ostracized." Some continue to hang around their mothers and sisters, clicking away for months unanswered. Eventually, though, they get the message. Fully grown males are solitary creatures. They approach a band of females—presumably not their immediate relatives—only in order to mate. To signal their arrival, they issue deep, booming sounds known as clangs. No one knows exactly what makes a courting sperm whale attractive to a potential mate; Gero told me that he had seen some clanging males greeted with great commotion and others with the cetacean equivalent of a shrug.

Female sperm whales, meanwhile, are exceptionally close. The adults in a unit not only travel and hunt together; they also appear to confer on major decisions. If there's a new mother in the group, the other members mind the calf while she dives for food. In some units, though not in Unit R, sperm whales even suckle one another's young. When a family is threatened, the adults cluster together to protect their offspring, and when things are calm the calves fool around.

"It's like my kids and their cousins," Gero said.

The day after I watched the successful drone flight, I went out with Gero to try to recover the recording device. More than twenty-four hours had passed, and it still hadn't been located. Gero decided to drive out along a peninsula called Scotts Head, at the southwestern tip of Dominica, where he thought he might be able to pick up the radio signal. As we wound around on the island's treacherously narrow roads, he described to me an idea he had for a children's book that, read in one direction, would recount a story about a human family that lives on a boat and looks down at the water and, read from the other direction, would be about a whale family that lives deep beneath the boat and looks up at the waves.

"For me, the most rewarding part about spending a lot of time in the culture of whales is finding these fundamental similarities, these fundamental patterns," he said. "And, you know, sure, they won't have a word for 'tree.' And there's some part of the sperm whale experience that our primate brain just won't understand. But those things that we share must be fundamentally important to why we're here."

After a while, we reached, quite literally, the end of the road. Beyond that was a hill that had to be climbed on foot. Gero was carrying a portable antenna, which he unfolded when we got to the top. If the recording unit had surfaced anywhere within twenty miles, Gero calculated, we should be able to detect the signal. It occurred to me that we were now trying to listen for a listening device. Gero held the antenna aloft and put his ear to some kind of receiver. He didn't hear anything, so, after admiring the view for a bit, we headed back down. Gero was hopeful that the device would eventually be recovered. But, as far as I know, it is still out there somewhere, adrift in the Caribbean.

THE FIRST SCIENTIFIC, or semi-scientific, study of sperm whales was a pamphlet published in 1835 by a Scottish ship doctor named Thomas Beale. Called *The Natural History of the Sperm Whale,* it proved so popular that Beale expanded the pamphlet into a book, which was issued under the same title four years later.

At the time, sperm whale hunting was a major industry, both in Britain and in the United States. The animals were particularly prized for their spermaceti, the waxy oil that fills their gigantic heads. Spermaceti is an excellent lubricant, and, burned in a lamp, produces a clean, bright light; in Beale's day, it could sell for five times as much as ordinary whale oil. (It is the resemblance between semen and spermaceti that accounts for the species' embarrassing name.)

Beale believed sperm whales to be silent. "It is well known among the most experienced whalers that they never produce any nasal or vocal sounds whatever, except a trifling hissing at the time of the expiration of the spout," he wrote. The whales, he said, were also gentle—"a most timid and inoffensive animal." Herman Melville relied heavily on Beale in composing *Moby-Dick.* (His personal copy of *The Natural History of the Sperm Whale* is now housed in Harvard's Houghton Library.) He attributed to sperm whales a "pyramidical silence."

"The whale has no voice," Melville wrote. "But then again," he went on, "what has the whale to say? Seldom have I known any profound being that had anything to say to this world, unless forced to stammer out something by way of getting a living."

The silence of the sperm whales went unchallenged until 1957. That year, two researchers from the Woods Hole Oceanographic Institution picked up sounds from a group they'd encountered off the coast of North Carolina. They detected strings of "sharp clicks," and speculated that these were made for the purpose of echolocation. Twenty years elapsed before one of the researchers, along with a different colleague from Woods Hole, determined that some sperm whale clicks were issued in distinctive, often repeated patterns, which the pair dubbed "codas." Codas seemed to be exchanged between whales and so, they reasoned, must serve some communicative function.

Since then, cetologists have spent thousands of hours listening to codas, trying to figure out what that function might be. Gero, who wrote his PhD thesis on vocal communication between sperm whales, told me that one of the "universal truths" about codas is their timing. There are always four seconds between the start of one coda and the beginning of the next. Roughly two of those seconds are given over to clicks; the rest is silence. Only after the pause, which may or may not be analogous to the pause a human speaker would put between words, does the clicking resume. Codas are clearly learned or, to use the term of art, socially transmitted. Whales in the eastern Pacific exchange one set of codas, those in the eastern Caribbean another, and those in the South Atlantic yet another. Baby sperm whales pick up the codas exchanged by their relatives, and before they can click them out proficiently they "babble."

The whales around Dominica have a repertoire of around twenty-five codas. These codas differ from one another in the number of their clicks and also in their rhythms. The coda known as three regular, or 3R, for example, consists of three clicks issued at equal intervals. The coda 7R consists of seven evenly spaced clicks. In seven increasing, or 7I, by contrast, the interval between the clicks grows longer; it's about five-hundredths of a second between the first two clicks, and between the last two it's twice that long.

In four decreasing, or 4D, there's a fifth of a second between the first two clicks and only a tenth of a second between the last two. Then there are syncopated codas. The coda most frequently issued by members of Unit R, which has been dubbed 1+1+3, has a cha-cha-esque rhythm and might be rendered in English as click . . . click . . . click-click-click.

If codas are in any way comparable to words, a repertoire of twenty-five represents a pretty limited vocabulary. But, just as no one can yet say what, if anything, codas mean to sperm whales, no one can say exactly what features are significant to them. It may be that there are nuances in, say, pacing or pitch that have so far escaped human detection. Already, CETI team members have identified a new kind of signal—a single click—that may serve as some kind of punctuation mark.

When whales are resting near the surface, their exchanges can last an hour or more. Even by human standards, sperm whale chatter is insistent and repetitive. "They're talking on top of each other all the time," Gero told me.

A snatch of dialogue recorded between two members of Unit R runs as follows. (Both Roger and Rita are adult females.)

Roger: 1+1+3
Rita: 1+1+3, 1+1+3
Roger: 9I
Rita: 1+1+3
Roger: 10I
Rita: 1+1+3, 1+1+3
Roger: 11I
Rita: 1+1+3
Roger: 10I, 11I, 1+1+3
Rita: 1+1+3

The "conversation" continues along much these same lines, until Rita finally changes her tune:

Rita: 1+1+3
Roger: 12R, 10I, 10I, 9I, 9I
Rita: 9I, 8I

NOT LONG AGO, suffering from writer's block, I asked ChatGPT if it could rewrite *Moby-Dick* from the whale's perspective. The chatbot began modestly. "As an AI language model, I can certainly attempt to rewrite a small excerpt," it told me. Then it launched into what it titled *Moby-Dick—The White Leviathan's Tale.*

In Melville's *Moby-Dick,* the plot turned on an unfortunate case of mistaken identity. Some other whale had gnawed off Ahab's leg; the white whale was as innocent as the driven snow.

"My heart was one of peace, my spirit one with the rhythm of the currents," ChatGPT wrote.

> I knew not of the vendetta forming against me, for I was but a creature of instinct, a guardian of the seas. But Captain Ahab's thirst for vengeance loomed large, and I became the object of his undying rage. The scar he bore from a previous encounter with my kind had branded me as his ultimate adversary. In his eyes, I represented a force of nature that defied his understanding, a living enigma of the ocean's might.

In paragraph seven, I was surprised—and excited—to see the whale/bot allude to the problem of interspecies communication:

> Deep within the abyss of the sea, I pondered the futility of the captain's quest. I longed to communicate with him, to show him that my kind, too, had emotions and families to protect. But the language barrier between our worlds remained an insurmountable chasm.

As anyone who has been conscious for the past ten months knows, ChatGPT is capable of amazing feats. It can write essays, compose sonnets, explain scientific concepts, and produce jokes (though these last are not necessarily funny). If you ask ChatGPT how it was created, it will tell you that first it was trained on a "massive corpus" of data from the internet. This phase consisted of what's called "unsupervised machine learning," which was performed by an intricate array of processing nodes known as a neural network. Basically, the "learning" involved filling in the blanks; according to ChatGPT, the exercise entailed "predicting the next word in a sentence given the context of the previous words." By digesting millions of web pages—and calculating and recalculating the odds—ChatGPT got so good at this guessing game that, without ever understanding English, it mastered the language. (Other languages it is "fluent" in include Chinese, Spanish, and French.)

In theory at least, what goes for English (and Chinese and Spanish and French) also goes for sperm whale. Provided that a computer model can be trained on enough data, it should be able to master coda prediction. It could then—once again in theory—generate sequences of codas that a sperm whale would find convincing. The model wouldn't understand sperm whale–ese, but it could, in a manner of speaking, speak it. Call it ClickGPT.

Currently, the largest collection of sperm whale codas is an archive assembled by Gero in his years on and off Dominica. The codas contain roughly a hundred thousand clicks. In a paper published last year, members of the CETI team estimated that, to fulfill its goals, the project would need to assemble some four billion clicks—which is to say, a collection roughly forty thousand times larger than Gero's.

"One of the key challenges toward the analysis of sperm whale (and more broadly, animal) communication using modern deep learning techniques is the need for sizable datasets," the team wrote.

In addition to bugging individual whales, CETI is planning to tether a series of three "listening stations" to the floor of the Caribbean Sea. The stations should be able to capture the codas of whales chatting up to twelve miles from shore. (Though inaudible above the waves, sperm whale clicks can register up to two hundred and thirty decibels, which is louder than a gunshot or a rock concert.) The information gathered by the stations will be less detailed than what the tags can provide, but it should be much more plentiful.

One afternoon, I drove with Gruber and CETI's station manager, Yaniv Aluma, a former Israeli Navy SEAL, to the port in Roseau, where pieces of the listening stations were being stored. The pieces were shaped like giant sink plugs and painted bright yellow. Gruber explained that the yellow plugs were buoys, and that the listening equipment— essentially, large collections of hydrophones—would dangle from the bottom of the buoys, on cables. The cables would be weighed down with old train wheels, which would anchor them to the seabed. A stack of wheels, rusted orange, stood nearby. Gruber suddenly turned to Aluma and, pointing to the pile, said, "You know, we're going to need more of these." Aluma nodded glumly.

The listening stations have been the source of nearly a year's worth of delays for CETI. The first was installed last summer, in water six thousand feet deep. Fish were attracted to the buoy, so the spot soon became popular among fishermen. After about a month, the fishermen noticed that the buoy was gone. Members of CETI's Dominica-based staff set out in the middle of the night on *CETI 1* to try to retrieve it. By the time they reached the buoy, it had drifted almost thirty miles offshore. Meanwhile, the hydrophone array, attached to the rusty train wheels, had dropped to the bottom of the sea.

The trouble was soon traced to the cable, which had been manufactured in Texas by a company that specializes in offshore oil-rig equipment. "They deal with infrastructure that's very solid," Aluma explained. "But a buoy has its own life. And they didn't calculate so well the torque or load on different motions—twisting and moving sideways." The company spent months figuring out why the cable had failed and finally thought it had solved the problem. In June, Aluma flew to Houston to watch a new cable go through stress tests. In the middle of the tests, the new design failed. To avoid further delays, the CETI team reconfigured the stations. One of the reconfigured units was installed late last month. If it doesn't float off, or in some other way malfunction, the plan is to get the two others in the water sometime this fall.

A SPERM WHALE'S head takes up nearly a third of its body; its narrow lower jaw seems borrowed from a different animal entirely; and its flippers are so small as to be almost dainty. (The formal name for the species is *Physeter macrocephalus,* which translates roughly as "big-headed blowhole.") "From just about any angle," Hal Whitehead, one of the world's leading sperm whale experts (and Gero's thesis adviser), has written, sperm whales appear "very strange." I wanted to see more of these strange-looking creatures than was visible from a catamaran, and so, on my last day in Dominica, I considered going on a commercial tour that offered customers a chance to swim with whales, assuming that any could be located. In the end—partly because I sensed that Gruber disapproved of the practice—I dropped the idea.

Instead, I joined the crew on *CETI 1* for what was supposed to be another round of drone tagging. After we'd been underway for about two hours, codas were picked up, to the northeast. We headed in that direction and soon came upon an extraordinary sight. There were at least ten whales right off the boat's starboard. They were all facing the same direction, and they were bunched tightly together, in rows. Gero identified them as members of Unit A. The members of Unit A were originally named for characters in Margaret Atwood novels, and they include Lady Oracle, Aurora, and Rounder, Lady Oracle's daughter.

Earlier that day, the crew on *CETI 2* had spotted pilot whales, or blackfish, which are known to harass sperm whales. "This looks very defensive," Gero said, referring to the formation.

Suddenly, someone yelled out, "Red!" A burst of scarlet spread through the water, like a great banner unfurling. No one knew what was going on. Had the pilot whales stealthily attacked? Was one of the whales in the group injured? The crowding increased until the whales were practically on top of one another.

Then a new head appeared among them. "Holy fucking shit!" Gruber exclaimed.

"Oh, my God!" Gero cried. He ran to the front of the boat, clutching his hair in amazement. "Oh, my God! Oh, my God!" The head belonged to a newborn calf, which was about twelve feet long and weighed maybe a ton. In all his years of studying sperm whales, Gero had never watched one being born. He wasn't sure anyone ever had.

As one, the whales made a turn toward the catamaran. They were so close I got a view of their huge, eerily faceless heads and pink lower jaws. They seemed oblivious of the boat, which was now in their way. One knocked into the hull, and the foredeck shuddered.

The adults kept pushing the calf around. Its mother and her relatives pressed in so close that the baby was almost lifted out of the water. Gero began to wonder whether something had gone wrong. By now, everyone, including the captain, had gathered on the bow. Pagani and another undergraduate, Aidan Kenny, had launched two drones and were filming the action from the air. Mevorach, meanwhile, was recording the whales through a hydrophone.

To everyone's relief, the baby began to swim on its own. Then the pilot whales showed up—dozens of them.

"I don't like the way they're moving," Gruber said.

"They're going to attack for sure," Gero said. The pilot whales' distinctive, wave-shaped fins slipped in and out of the water.

What followed was something out of a marine mammal *Lord of the Rings*. Several of the pilot whales stole in among the sperm whales. All that could be seen from the boat was a great deal of thrashing around. Out of nowhere, more than forty Fraser's dolphins arrived on the scene. Had they come to participate in the melee or just to rubberneck? It was impossible to tell. They were smaller and thinner than the pilot whales (which, their name notwithstanding, are also technically dolphins).

"I have no prior knowledge upon which to predict what happens next," Gero announced. After several minutes, the pilot whales retreated. The dolphins curled through the waves. The sperm whales remained bunched together. Calm reigned. Then the pilot whales made another run at the sperm whales. The water bubbled and churned.

"The pilot whales are just being pilot whales," Gero observed. Clearly, though, in the great "struggle for existence," everyone on board *CETI 1* was on the side of the baby.

The skirmishing continued. The pilot whales retreated, then closed in again. The drones began to run out of power. Pagani and Kenny piloted them back to the catamaran to exchange the batteries. These were so hot they had to be put in the boat's refrigerator. At one point, Gero thought that he spied the new calf, still alive and well. (He would later, from the drone footage, identify the baby's mother as Rounder.) "So that's good news," he called out.

The pilot whales hung around for more than two hours. Then, all at once, they were gone. The dolphins, too, swam off. "There will never be a day like this again," Gero said as *CETI 1* headed back to shore.

That evening, everyone who'd been on board *CETI 1* and *CETI 2* gathered at a dockside restaurant for a dinner in honor of the new calf. Gruber made a toast. He thanked the team for all its hard work. "Let's hope we can learn the language with that baby whale," he said.

I was sitting with Gruber and Gero at the end of a long table. In between drinks, Gruber suggested that what we had witnessed might not have been an attack. The scene, he proposed, had been more like the last act of *The Lion King,* when the beasts of the jungle gather to welcome the new cub.

"Three different marine mammals came together to celebrate and protect the birth of an animal with a sixteen-month gestation period," he said. Perhaps, he hypothesized, this was a survival tactic that had evolved to protect mammalian young against sharks, which would have been attracted by so much blood and which, he pointed out, would have been much more numerous before humans began killing them off.

"You mean the baby whale was being protected by the pilot whales from the sharks that aren't here?" Gero asked. He said he didn't even know what it would mean to test such a theory. Gruber said they could look at the drone footage and see if the sperm whales had ever let the pilot whales near the newborn and, if so, how the pilot whales had responded. I couldn't tell whether he was kidding or not.

"That's a nice story," Mevorach interjected.

"I just like to throw ideas out there," Gruber said.

THE COMPUTER SCIENCE and Artificial Intelligence Laboratory (CSAIL), at MIT, occupies a Frank Gehry–designed building that appears perpetually on the verge of collapse. Some wings tilt at odd angles; others seem about to split in two. In the lobby of the building, there's a vending machine that sells electrical cords and another that dispenses caffeinated beverages from around the world. There's also a yellow sign of the sort you might see in front of an elementary school. It shows a figure wearing a backpack and carrying a briefcase and says "NERD XING."

Daniela Rus, the head of CSAIL (pronounced "see-sale"), is a roboticist. "There's such a crazy conversation these days about machines," she told me. We were sitting in her office, which is dominated by a robot, named Domo, who sits in a glass case. Domo has a metal torso and oversized, goggly eyes. "It's either machines are going to take us down or machines are going to solve all of our problems. And neither is correct."

Along with several other researchers at CSAIL, Rus has been thinking about how CETI might eventually push beyond coda prediction to something approaching coda comprehension. This is a formidable challenge. Whales in a unit often chatter before they dive. But what are they chattering about? How deep to go, or who should mind the calves, or something that has no analogue in human experience?

"We are trying to correlate behavior with vocalization," Rus told me. "Then we can begin to get evidence for the meaning of some of the vocalizations they make."

She took me down to her lab, where several graduate students were tinkering in a thicket of electronic equipment. In one corner was a transparent plastic tube loaded with circuitry, attached to two white plastic flippers. The setup, Rus explained, was the skeleton of a robotic turtle. Lying on the ground was the turtle's plastic shell. One of the students hit a switch and the flippers made a paddling motion. Another student brought out a two-foot-long robotic fish. Both the fish and the turtle could be configured to carry all sorts of sensors, including underwater cameras.

"We need new methods for collecting data," Rus said. "We need ways to get close to the whales, and so we've been talking a lot about putting the sea turtle or the fish in water next to the whales, so that we can image what we cannot see."

CSAIL is an enormous operation, with more than fifteen hundred staff members and students. "People here are kind of audacious," Rus said. "They really love the wild and crazy ideas that make a difference." She told me about a diver she had met who had swum with the sperm whales off Dominica and, by his account at least, had befriended one. The whale seemed to like to imitate the diver; for example, when he hung in the water vertically, it did too.

"The question I've been asking myself is: Suppose that we set up experiments where we engage the whales in physical mimicry," Rus said. "Can we then get them to vocalize while doing a motion? So, can we get them to say, 'I'm going up'? Or can we get them to say, 'I'm hovering'? I think that, if we were to find a few snippets of vocalizations that we could associate with some meaning, that would help us get deeper into their conversational structure."

While we were talking, another CSAIL professor and CETI collaborator, Jacob Andreas, showed up. Andreas, a computer scientist who works on language processing, said that he had been introduced to the whale project at a faculty retreat. "I gave a talk about understanding neural networks as a weird translation problem," he recalled. "And Daniela came up to me afterwards and she said, 'Oh, you like weird translation problems? Here's a weird translation problem.'"

Andreas told me that CETI had already made significant strides, just by reanalyzing Gero's archive. Not only had the team uncovered the new kind of signal, but also it had found that codas have much more internal structure than had previously been recognized. "The amount of information that this system can carry is much bigger," he said.

"The holy grail here—the thing that separates human language from all other animal communication systems—is what's called 'duality of patterning,'" Andreas went on. "Duality of patterning" refers to the way that meaningless units—in English, sounds like "sp" or "ot"—can be combined to form meaningful units, like "spot." If, as is suspected, clicks are empty of significance but codas refer to something, then sperm whales, too, would have arrived at duality of patterning. "Based on what we know about how the coda inventory works, I'm optimistic—though still not sure—that this is going to be something that we find in sperm whales," Andreas said.

THE QUESTION OF whether any species possesses a "communication system" comparable to that of humans is an open and much debated one. In the nineteen-fifties, the behaviorist B. F. Skinner argued that children learn language through positive reinforcement; therefore, other animals should be able to do the same. The linguist Noam Chomsky had a different view. He dismissed the notion that kids acquire language via conditioning, and also the possibility that language was available to other species.

In the early nineteen-seventies, a student of Skinner's, Herbert Terrace, set out to confirm his mentor's theory. Terrace, at that point a professor of psychology at Columbia, adopted a chimpanzee, whom he named, tauntingly, Nim Chimpsky. From the age of two weeks, Nim was raised by people and taught American Sign Language. Nim's interactions with his caregivers were videotaped, so that Terrace would have an objective record of the chimp's progress. By the time Nim was three years old, he had a repertoire of eighty signs and, significantly, often produced them in sequences, such as "banana me eat banana" or "tickle me Nim play." Terrace set out to write a book about how Nim had crossed the language barrier and, in so doing, had made a monkey of his namesake. But then Terrace double-checked some details of his account against the tapes. When he looked carefully at the videos, he was appalled. Nim hadn't really learned ASL; he had just learned to imitate the last signs his teachers had made to him.

"The very tapes I planned to use to document Nim's ability to sign provided decisive evidence that I had vastly over-estimated his linguistic competence," Terrace wrote.

Since Nim, many further efforts have been made to prove that different species—orangutans, bonobos, parrots, dolphins—have a capacity for language. Several of the animals who were the focus of these efforts—Koko the gorilla, Alex the gray parrot—became international celebrities. But most linguists still believe that the only species that possesses language is our own.

Language is "a uniquely human faculty" that is "part of the biological nature of our species," Stephen R. Anderson, a professor emeritus at Yale and a former president of the Linguistic Society of America, writes in his book *Doctor Dolittle's Delusion*.

Whether sperm whale codas could challenge this belief is an issue that just about everyone I talked to on the CETI team said they'd rather not talk about.

"Linguists like Chomsky are very opinionated," Michael Bronstein, the Oxford professor, told me. "For a computer scientist, usually a language is some formal system, and often we talk about artificial languages." Sperm whale codas "might not be as expressive as human language," he continued. "But I think whether to call it 'language' or not is more of a formal question."

"Ironically, it's a semantic debate about the meaning of language," Gero observed.

Of course, the advent of ChatGPT further complicates the debate. Once a set of algorithms can rewrite a novel, what counts as "linguistic competence"? And who—or what—gets to decide?

"When we say that we're going to succeed in translating whale communication, what do we mean?" Shafi Goldwasser, the Radcliffe Institute fellow who first proposed the idea that led to CETI, asked.

"Everybody's talking these days about these generative AI models like ChatGPT," Goldwasser, who now directs the Simons Institute for the Theory of Computing, at the University of California, Berkeley, went on. "What are they doing? You are giving them questions or prompts, and then they give you answers, and the way that they do that is by predicting how to complete sentences or what the next word would be. So you could say that's a goal for CETI—that you don't necessarily understand what the whales are saying, but that you could predict it with good success. And, therefore, you could maybe generate a conversation that would be understood by a whale, but maybe you don't understand it. So that's kind of a weird success."

Prediction, Goldwasser said, would mean "we've realized what the pattern of their speech is. It's not satisfactory, but it's something."

"What about the goal of understanding?" she added. "Even on that, I am not a pessimist."

THERE ARE NOW an estimated eight hundred and fifty thousand sperm whales diving the world's oceans. This is down from an estimated two million in the days before the species was commercially hunted. It's often suggested that the darkest period for *P. macrocephalus* was the middle of the nineteenth century, when Melville shipped out of New Bedford on the *Acushnet*. In fact, the bulk of the slaughter took place in the middle of the twentieth century, when sperm whales were pursued by diesel-powered ships the size of factories. In the eighteen-forties, at the height of open-boat whaling, some five thousand sperm whales were killed each year; in the nineteen-sixties, the number was six times as high. Sperm whales were boiled down to make margarine, cattle feed, and glue. As recently as the nineteen-seventies, General Motors used spermaceti in its transmission fluid.

Near the peak of industrial whaling, a biologist named Roger Payne heard a radio report that changed his life and, with it, the lives of the world's remaining cetaceans. The report noted that a whale had washed up on a beach not far from where Payne was working, at Tufts University. Payne, who'd been researching moths, drove out to see it. He was so moved by the dead animal that he switched the focus of his research. His investigations led him to a naval engineer who, while listening for Soviet submarines, had recorded eerie underwater sounds that he attributed to humpback whales. Payne spent years studying the recordings; the sounds, he decided, were so beautiful and so intricately constructed that they deserved to be called "songs." In 1970, he arranged to have *Songs of the Humpback Whale* released as an LP.

"I just thought: the world has to hear this," he would later recall. The album sold briskly, was sampled by popular musicians like Judy Collins, and helped launch the Save the Whales movement. In 1979, *National Geographic* issued a "flexi disc" version of the songs, which it distributed as an insert in more than ten million copies of the magazine. Three years later, the International Whaling Commission declared a "moratorium" on commercial hunts that remains in effect today. The move is credited with having rescued several species, including humpbacks and fin whales, from extinction.

Payne, who died in June at the age of eighty-eight, was an early and ardent member of the CETI team. (This was the case, Gruber told me, even though he was disappointed that the project was focusing on sperm whales rather than on humpbacks, which, he maintained, were more intelligent.) Just a few days before his death, Payne published an op-ed piece explaining why he thought CETI was so important.

Whales, along with just about every other creature on earth, are now facing grave new threats, he observed, among them climate change. How to motivate "ourselves and our fellow humans" to combat these threats?

"Inspiration is the key," Payne wrote. "If we could communicate with animals, ask them questions and receive answers—no matter how simple those questions and answers might turn out to be—the world might soon be moved enough to at least start the process of halting our runaway destruction of life."

Several other CETI team members made a similar point. "One important thing that I hope will be an outcome of this project has to do with how we see life on land and in the oceans," Bronstein said. "If we understand—or we have evidence, and very clear evidence in the form of language-like communication—that intelligent creatures are living there and that we are destroying them, that could change the way that we approach our Earth."

"I always look to Roger's work as a guiding star," Gruber told me. "The way that he promoted the songs and did the science led to an environmental movement that saved whale species from extinction. And he thought that CETI could be much more impactful. If we could understand what they're saying, instead of 'save the whales' it will be 'saved by the whales.'

"This project is kind of an offering," he went on. "Can technology draw us closer to nature? Can we use all this amazing tech we've invented for positive purposes?"

ChatGPT shares this hope. Or at least the AI-powered language model is shrewd enough to articulate it. In the version of *Moby-Dick* written by algorithms in the voice of a whale, the story ends with a somewhat ponderous but not unaffecting plea for mutuality: "I, the White Leviathan, could only wonder if there would ever come a day when man and whale would understand each other, finding harmony in the vastness of the ocean's embrace."

Originally published in *The New Yorker*,
September 11, 2023.

LIFE ON A LITTLE-KNOWN PLANET

Profile of Entomologist David Wagner,
Who Is Racing to Find Caterpillars
Before They Disappear

THE DEVILS RIVER, in southwestern Texas, runs, mirage-like, along the edge of the Chihuahuan Desert, through some of the most barren countryside in the United States. Access to the river is limited; unless you're in a kayak, the only way to travel upstream is along a skein of rutted dirt roads. It was on one of these roads that, a few years ago, David Wagner noticed a shrub that seemed to him peculiarly filled with promise.

Wagner is an entomologist who teaches at the University of Connecticut. He has close-cropped silvery hair and a square jaw and bears a passing resemblance to George C. Scott playing General Buck Turgidson. The way other people might recall a marvelous restaurant or a heartbreaking vista, Wagner remembers a propitious plant. He has friends who own a house along the Devils River, and each time he has visited them he has stopped by the exact same shrub to investigate. No luck. This past October, I was traveling with him when he tried yet again. He spread a white nylon sheet on the ground, then started whacking the bush with a pole to dislodge anything that might be clinging to it.

"Un-fucking-believable!" he exclaimed. I was whacking a plant nearby, just for the hell of it. Wagner held out his hand. A caterpillar about three-quarters of an inch long was wriggling across his palm. It looked brownish and totally ordinary until I examined it under a loupe, at which point it was revealed to be flamboyantly striped, with yellow and red splotches and two black, hornlike protuberances stick-ing out of its back. Based on a series of taxonomic calculations, Wag-ner was convinced that the caterpillar was the juvenile form of an exceptionally rare moth known as *Ursia furtiva*.

"No one's ever seen this before," he told me. If Wagner was the first person to lay eyes on an *U. furtiva* caterpillar, I figured, that meant I was the second. Un-fucking-believable.

Caterpillars are to lepidoptera—butterflies and moths—what grubs are to beetles and maggots are to flies; they are larvae. Even among nature lovers, larvae tend to be unloved. For every ten butterfly fanci-ers, there are approximately zero caterpillar enthusiasts. The reason for this will, to most, seem obvious. The worm in the apple is usually a caterpillar.

Wagner specializes in caterpillars, or, it might be more accurate to say, is consumed by them. (They are, he suggested to me, the reason he is no longer married.) Probably he knows more about the caterpillars of the U.S. than anyone else in the country, and possibly he knows more about caterpillars in general than anyone else on the planet. When he travels, it's not uncommon for him to return home with a suitcase full of specimens. Most of these he has injected with alcohol; some, though, may remain alive, nestled in little vials of their favorite plants.

Wagner's *Caterpillars of Eastern North America,* published in 2005, runs to nearly five hundred pages. It relates the life histories of roughly that many species and is considered the definitive field guide on the subject. Wagner is now thirteen years into an even more ambitious project, *Caterpillars of Western North America,* which he plans to pub-lish in four volumes.

The implicit argument of Wagner's work is that every larva mat-ters, no matter how small, squishy, and unassuming. Each new spe-cies that he collects is a different answer to life's great conundrum: how to survive on planet Earth. Each has a unique and often startling story to tell.

"I want to write each species' account so you want to read one more," he told me. "I want it to be a page-turner."

• • •

WAGNER, WHO IS SIXTY-SIX, grew up pretty much all over the place. His father, a metallurgist, worked for U.S. Steel, and the family moved whenever he was assigned to a new project—a bridge in one state, a pipeline in another. In second grade, Wagner attended three different schools. "I don't know where I'm from," he told me. At one school, in Missouri, he remembers, he was greeted by kids throwing rocks at him.

From an early age, he was interested in bugs. He collected them in an old cigar box, which moved from town to town with him. "They meant the world to me," he said. "I would go into my room at night, and I would look at them with a hand lens. There was an infinite amount of beauty and complexity there." Though his parents didn't share his interest, they abetted it. They bought him *A Field Guide to the Butterflies of North America, East of the Great Plains,* which he read so often that the cover fell off.

"That book was a portal for me," Wagner said. "I was able to see into another world, and I was fascinated by it." A Christmas letter his father sent to a relative in 1966 describes Wagner, then in fifth grade, as interested in "anything that crawls or moves." His father writes that he is sorry to be unable to answer Wagner's "million and one questions per day."

In college, at Colorado State, Wagner took classes from Howard Ensign Evans, an expert on wasps who had given up a tenured professorship at Harvard to move to the Rockies. Evans was the author of *Life on a Little-Known Planet,* a book that, by entomological standards, counted as a blockbuster. Published in 1968, it was a paean to the sorts of insects people usually regard as pests—locusts, for example, and bedbugs. (Evans dedicated the book to the lice and silverfish that shared his office.) It was also a plea for insect conservation. Americans, Evans lamented, seemed more curious about what might live on Mars than about the many uncataloged creatures living right beneath their feet.

"Is it sensible to poke about for strange beings in space while we blindly exterminate those about us?" he asked. "It is said that not much more than half the organisms on earth have yet been described. As one who has several times discovered insects new to science in his own back yard, I can believe this."

In graduate school, at the University of California, Berkeley, Wagner devoted himself to ghost moths. Holdovers from the days of the dinosaurs, ghost moths exhibit many curious behaviors. "When they fly, their wings flap around independently, which is a really ancient, uncoordinated mode of flight," he explained. Males of some ghost moth species perform a sort of aerial cha-cha as a prelude to sex. Others have specialized legs loaded with come-hither chemicals. Much remains to be learned about ghost moths, and Wagner would probably have kept right on studying them had it not been for an ecological disaster.

Back in the eighteen-sixties, in a bungled attempt to establish a silk business, a Frenchman named Étienne Léopold Trouvelot imported gypsy moths from Europe to Massachusetts. Some of Trouvelot's moths, which are now less offensively referred to as spongy moths, got loose. Their eggs hatched into spongy moth caterpillars, which proceeded to defoliate much of New England. By the nineteen-nineties, spongy moths had pushed as far as Virginia, and their caterpillars were wreaking havoc in Shenandoah National Park. The Forest Service wanted to strike back with insecticide, but there was concern that a lot of other moths and butterflies would be killed off in the process. Researchers set up experimental plots in the Blue Ridge Mountains, sprayed some of them, and left the others alone. Then they gathered from the plots all the caterpillars they could find and sent them to Wagner at UConn's main campus, in Storrs.

"The first week, they sent six thousand caterpillars," Wagner recalled. During the next couple of weeks, more shipments arrived, until the caterpillars numbered thirteen thousand.

As anyone who has ever witnessed a spongy moth infestation or just read Eric Carle knows, caterpillars eat voraciously. Many are picky, though, and will consume only certain plants. Wagner scrambled to staff his lepidopteran nursery. "If you could push a shopping cart, you could get a job," he recalled.

Among the thousands of larvae sent by the Forest Service were lots that Wagner didn't recognize. He called around to various colleagues, who weren't much help. At that point, there was no definitive guide to the caterpillars of eastern North America to consult. Just to have a way to refer to their charges, Wagner and his assistants started making up nicknames.

A particularly colorful caterpillar became known as the Dazzler. Another, with a groove on the tip of its rearmost segment, they dubbed Plumber's Butt. For Wagner, the experience was eye-opening. Every butterfly or moth was once a larva. And yet there was a lack of basic information on caterpillars—what they looked like, what they ate, how they got through the winter. Taxonomically speaking, they were an undiscovered country.

IT WAS ABOUT TEN A.M. when Wagner and I left the bush by the Devils River and headed northwest, deeper into the Chihuahuan Desert. It had rained a few weeks earlier—this was the reason Wagner was making the trip—and the foliage was surprisingly green. Every so often, Wagner would spot an interesting plant and screech off the highway. He would pull out his white nylon beating sheet, which was stretched across a frame, like a kite, and thump away at the vegetation. Anything he liked the looks of went into a plastic vial, and then into the canvas collecting bag he wore slung over his shoulder. With all the stopping and starting, we didn't cover much distance, even though, when we were driving, it was often at more than ninety miles an hour. Finally, at almost nine p.m., we pulled into the town of Alpine, home of Sul Ross State University, where Wagner had an appointment scheduled for the following morning.

Wagner likes to say "It takes a village," by which he means that no one person, no matter how ardent, can beat every bush. During the last decade, he has built up—or, you might say, collected—an extensive network of helpers. Some have professional training in entomology; others are amateurs he has persuaded to keep an eye out for strange-looking larvae. Still others own ranches where rare species might lurk, or know people who do. Not infrequently, his collaborators FedEx him their discoveries.

I ended up meeting—and, in some cases, staying in the homes of—several members of Wagner's Texas network. They included his friends by the Devils River, Tracy and Dave Barker, both herpetologists.

"Dave's good at enlisting people," Tracy told me, referring to Wagner. "He gets to these people, and then all of a sudden they have all these jars in their kitchens."

"Dave is the one person I know who will always get back to you," Delmar Cain, a retired lawyer, told me. "Plus, he will give you fifteen jobs in the meantime—things that you need to be on the lookout for."

"Dave has contaminated everybody," a third member of the network, Michael Powell, a West Texas botanist, said, sighing.

On the way to Alpine, Wagner had told me that he was "courting" a friend of Powell's named Kelsey Wogan, by which he meant nothing romantic. (As a gift for her, he had brought along an extra beating sheet.) We met up with Wogan in the herbarium on the Sul Ross campus. She was carrying a basket filled with ziplock bags. Wagner opened the first ziplock and fished out a caterpillar.

"I've never seen anything like this before!" he exclaimed, reaching for the loupe that dangled from a cord around his neck.

To the naked eye, the caterpillar appeared gray with black splotches. Viewed at ten times magnification, it was twenty times weirder. Instead of gray, it was a dusty rose, with stripes of salmon dots that ran along its sides and met up above its many eyes. (Most caterpillars have twelve eyes, six on each side.) The black splotches were raised, like tiny hillocks, and covered in even tinier white stipples.

"This is super cool!" Wagner said. He didn't even know what genus the caterpillar belonged to, which meant that it, too, might prove to be some kind of first.

Wogan passed him another plastic bag. The caterpillar in this one was, relatively speaking, gigantic—nearly the size of a cocktail sausage. It was a velvety bronze speckled with turquoise. To me, it seemed even cooler than the first, but Wagner immediately recognized it as the caterpillar of a common butterfly, the two-tailed swallowtail, which made it scientifically uninteresting. Perhaps sensing my disappointment, he pinched the caterpillar's thorax. Two tentacles emerged from a hidden compartment, along with some goo that smelled like vomit. This, he explained, was a defensive strategy—a trick the creature used to ward off birds. Wogan seemed suitably disgusted. She handed over more bags, filled with more caterpillars.

FROM A CATERPILLAR'S PERSPECTIVE, humans are boring. The young they squeeze out of their bodies are just miniature versions of themselves, with all the limbs and appendages they'll ever have. As they mature, babies get bigger and stronger and hairier, but that's about it. Caterpillars, for their part, are continually reinventing themselves. They emerge from tiny, jewel-like eggs and for their first meal often eat their own egg cases. Once they reach a certain size, they sprout a second head, just behind the first. They then wriggle free of their old skin, the way a diver might wriggle out of a wetsuit. (In the process, the old head drops off.) In the course of their development, they will complete this exercise three, four, in some species sixteen times, often trying out a new look along the way. The spicebush swallowtail, for example, which is found throughout the eastern U.S., emerges from its egg mottled in black and white. This color scheme allows it to pass itself off as a bird dropping. After its third molt, as a so-called fourth instar, it turns green (or brown), with two yellow-and-black spots on its head. The spots, which look uncannily like a pair of eyes, enable the swallowtail to pretend it's a snake.

After running through its allotment of instars, a caterpillar ceases to be itself and becomes a pupa. It sheds its skin one last time and develops a hardened shell. Inside this shell, its body dissolves. Then, from bundles of cells known as imaginal disks, a new body takes form. Some disks develop into legs, some wings, some genitalia, and so on. The creature that emerges retains almost nothing of its juvenile self except, weirdly, its memories.

As a way of life, this radical, whole-body transformation is ancient. It arose some three hundred and fifty million years ago, during the Carboniferous period. How, exactly, the process evolved is still debated, but it has proved wildly popular. Not only moths and butterflies undergo complete metamorphosis; so, too, do beetles, flies, wasps, fleas, and lacewings.

WHEN WE'D FIRST set out on our collecting expedition, Wagner had issued a warning. There wasn't going to be much time on the trip for sleeping or eating. "We can do two meals a day," he had said. "But I don't think we'll ever see three." In fact, as we tore around West Texas, we hadn't skipped many meals. But Wagner, at least, was staying up later and later. The more caterpillars he collected, the longer it took to care for them. One night in Alpine, I volunteered to help. By this point, we were five days out, and Wagner's traveling menagerie had grown to include some seventy hungry caterpillars. His plastic vials were arranged in rows on the nightstand of his hotel room. He pulled several bagfuls of leaves from the room's mini-fridge.

Any animal that eats voraciously poops voluminously. The first task of the evening involved cleaning the shit—or, more politely, frass—out of the vials, a practice Wagner referred to as "mucking the stalls." He handed me a paper plate. I was to dump the contents of a vial onto the plate, swab out the container, and then put the caterpillar back inside, along with a fresh sprig of its host plant. As we were mucking the stalls, Wagner realized that one of the caterpillars Wogan had found had gone missing. It was another cocktail-sausage-size creature, midnight black in color. We searched around on the floor for a while but couldn't find it. Then I stood up to fetch something and felt a hideous squish underfoot.

"The caterpillar gods give and they taketh away," Wagner pronounced, wiping the black gunk off the rug. He tried to console me by claiming the caterpillar had been a relatively common one.

On top of all the mucking and feeding, each addition to his collection had to be logged, with a note made of precisely where it had been found and on what plant. Then—most time-consuming of all—it had to be photographed. Moths and butterflies make handsome specimens that can retain their looks for centuries. Caterpillars, sadly, don't last. Pickled in alcohol, they become soggy and discolored; unpickled, they rot. Photographs are pretty much the only way to preserve them.

Wagner treats his subjects like so many many-legged Christy Turlingtons. He poses them on sprigs that he attaches to a little stand, which he arranges in front of a green background. Then he leans in with a huge macro lens. Wagner confessed to me that he wants his caterpillars to be hungry during a photo shoot; that way, he can catch them in what's called their resting posture. (At rest, a caterpillar colored like a twig will stick out at an angle, just as a real twig would do.) Caterpillars don't see very well; probably they can detect light and dark, but not much else. Nevertheless, Wagner likes to capture them "looking" at the camera.

"Eye contact is critical," he told me. "People are going to connect to that."

The night of the squish, I hung around for a while to watch Wagner shoot and reshoot that day's finds. He was wearing his usual field uniform—jeans, sneakers, a short-sleeved button-down shirt, and a cap printed with the logo of Caterpillar, Inc. (The cap, he told me, was "tongue-in-cheek.") He was so much bigger than his models that when I took a picture of him taking a picture of them it looked like a man photographing nothing.

Seen in Wagner's extreme-closeup shots, the caterpillars were, once again, spectacular. One, which Wagner had found on a pine tree, was gray, with uneven dark patches that looked like shadows and made the caterpillar appear to have the texture of pine bark. Another, which he had found on an oak leaf, was dark green with a white stripe down its back that mimicked the leaf's midrib, and light-green stripes that mimicked its veins. "I think a lot of people find insects repugnant or ugly," Wagner told me. "I can't see that."

Around midnight, I decided it was time to turn in. Wagner, I learned the next morning, had stayed up for three more hours. "The only reason I go to bed is so I don't mess up the next day," he said.

. . .

THERE ARE ROUGHLY sixty-five hundred species of mammals, nine thousand species of amphibians, and eleven thousand species of birds. These are what people tend to think of when they picture the world's biodiversity. But the planet's real diversity lies mostly beneath our regard. The largest family of beetles, the Curculionidae, commonly known as weevils, contains some sixty thousand described species; another beetle family, the Tenebrionidae, comprises twenty thousand species. It is estimated that in one family of parasitic wasps, the Ichneumonidae, there are nearly a hundred thousand species, which is more than there are of vertebrates of all kinds. (Ichneumonids inject their eggs into the larvae of other insects. Darwin adduced their existence as a powerful argument against intelligent design; he could not, he wrote, imagine a "beneficent and omnipotent God" purposely creating such a fiendish creature.) There are, in fact, so many insect species—at least two million and possibly as many as ten million—that Robert May, an Australian physicist turned theoretical ecologist, once joked, "To a good approximation, all species are insects!"

In keeping with their variety, insects play a vital role in virtually every terrestrial ecosystem. Roughly three-quarters of the world's flowering plants depend on insects for pollination. Insects are also crucial seed dispersers; many plants stud their seeds with tiny treats to entice ants to carry them off. And they're key decomposers. (When a person dies, blowflies arrive on the scene within minutes; in warm weather, blowfly maggots can eat through most of a corpse within a week.)

Legions of other creatures, meanwhile, depend on insects for food. Insectivorous mammals include hedgehogs, shrews, and most species of bats. Just about all amphibians consume insects, as do many species of reptiles and freshwater fish. Lots of birds rely on insects, particularly during breeding season: before they fledge, a clutch of young chickadees will consume as many as six thousand caterpillars. Collectively, insects transfer more energy from plants to animals than any other group. They are the solder that holds food chains together. This vital work is, at least by *Homo sapiens,* underappreciated. To the extent that we attend to insects, usually it's to those that irk us. If there's a moral to Wagner's work, it is that, instead of arrogantly blundering along, we ought to pause and look more closely. What we'd discover is one marvel after another.

The caterpillar of the silver-spotted skipper, I learned from *Caterpillars of Eastern North America,* uses an air gun–like appendage in its anus to send its frass pellets soaring. This practice, known as "fecal firing," discombobulates parasitic wasps. The silvery blue caterpillar possesses a "nectary organ" that dispenses a sugary liquid; ants attracted to the liquid are enlisted as bodyguards. The camouflaged looper confuses potential predators by chewing off bits of plant matter, like petals, and attaching them to its back. When threatened, the catalpa sphinx caterpillar spews out green goo and thrashes around violently. The walnut sphinx caterpillar, too, is a thrasher; instead of spitting up goo, it whistles through its air holes, or spiracles. The lace-capped caterpillar is colored to look like a piece of dying vegetation; when it eats out a section of a leaf, it fills the gap with its body, in effect becoming the damage it has caused.

One day, toward the end of our collecting trip, Wagner and I pulled off the road a few miles from the Mexican border. As usual, Wagner laid out a beating sheet and started to whack at the vegetation. By this point, I was getting to be a pretty experienced whacker myself. I was also getting better at distinguishing caterpillars from the bits of plant debris that fell onto the beating sheet Wagner had loaned me. (The key difference is that caterpillars move.) After laying into a plant called devil's claw, I found a small, unexceptional-looking green caterpillar, which, viewed under a hand lens, still appeared to me to be small and green and unexceptional. I showed it to Wagner, who immediately noticed something special about it. The caterpillar was covered in minute spines, but, he explained, it lacked the markings common to similarly spiny caterpillars from the genus *Heliothis.*

"I think your little green thing may be significant," he told me. He put it in a vial and said that he was going to send it off for genetic analysis. Depending on the results, it could get its own entry in *Caterpillars of Western North America.*

• • •

AROUND THE TIME that Wagner turned sixty, in 2016, he started to think about retiring and, possibly, moving out West. But then another, much bigger ecological crisis intervened—what's become known, perhaps prematurely, perhaps not, as the insect apocalypse.

The first sign of the "apocalypse" came from the city of Krefeld, in western Germany. For decades, members of the Krefeld Entomological Society had been trapping insects in protected areas near the city. Every bug they caught they weighed and preserved in bottles of alcohol.

In 2013, society members returned to two sites they had first sampled in 1989. To their surprise, the total mass of the insects they caught was just a fraction of what it had been the first time around. When they resampled other areas, the results were much the same. They passed on their data to a team of scientists, who wrote up their conclusions in a paper that ran in *PLOS One* in 2017. In less than thirty years, "total flying insect biomass" in the areas sampled had dropped by three-quarters, the paper said—suggesting that the entire flying-insect community had been "decimated."

Other papers filled with equally bleak statistics soon followed. A study of the Upper Mississippi River and the Western Lake Erie Basin found that the number of mayflies emerging from the two waterways had dropped by more than half just since 2012. (Mayflies can form such large swarms that they are visible on radar.) An analysis of data collected each summer in Ohio showed that butterfly sightings in the state had declined by a third in little more than two decades. Researchers working in the Hubbard Brook Experimental Forest, in New Hampshire's White Mountains, discovered that the number of beetles in the forest had fallen by more than 80 percent since the mid-nineteen-seventies, and that some beetle families had disappeared entirely.

As the papers piled up, a countermovement took shape. Some researchers argued that there was a bias toward doom. Studies that found no particular trend in insect numbers were less interesting than those that suggested a crisis and therefore were less likely to get published. Other researchers pushed back. Even if insects were doing okay in some places, they said, the outlines of the problem were clear and the stakes too high to wait.

"If we do not take action now to address declines in insect abundance and diversity, we will very likely face problems . . . that will make many previous challenges faced by human civilization seem tame by comparison," a trio of researchers led by Matt Forister, of the University of Nevada, Reno, wrote.

From his own experience, Wagner knew that many species that used to be common had become rare. He had published a paper in 2012 titled "Moth Decline in the Northeastern United States"; it noted that several large, showy moths, like the promethea silkmoth, which had been easy to find when he arrived at UConn, in the late nineteen-eighties, had since vanished. The more data that came in the more concerned he became.

In the fall of 2019, Wagner organized a session called "Insect Decline in the Anthropocene" for the annual meeting of the Entomological Society of America, which was held that year in St. Louis. It turned out to be one of the conference's best-attended sessions. Most of the speakers, including two of the authors of the Krefeld paper, offered dire warnings; a few, though, warned against being too dire.

I first met Wagner not long after that session, at a Christmas party where I was also introduced to several giant cockroaches. The party was held in the entomology offices of the American Museum of Natural History. That evening, as we snacked on fried crickets, Wagner told me, "For the first time, I think people are really worried about ecosystem services and all the things insects do to sustain the planet."

LAST MONTH, I met up with Wagner again at the Natural History Museum. He had received a grant from the National Science Foundation to organize a five-year research program on insect decline, and the museum was hosting a conference to kick off the effort. As if to underscore the urgency of the situation, on the first day of the meeting a new paper on the status of insects appeared. The paper, by researchers at the Chinese Academy of Agricultural Sciences, found that the number of flying insects using a key migration corridor between China and the Korean Peninsula had dropped significantly in just over fifteen years, and that the losses were higher among species that prey on other insects. It suggested that food webs were starting to break down. "Active conservation of insect communities is pivotal," the study concluded.

Much of the conference took place in a wood-paneled room presided over by a portrait of Teddy Roosevelt. Participants included experts on, among many other groups, bees, dragonflies, katydids, and dung beetles. Wagner opened the gathering by saying he was there "because the bugs demanded that I do this."

Some of the presentations were held over Zoom, with all the attendant technical glitches. Matt Forister talked about declining butterfly counts in California's Central Valley and, even more worrying, at high elevations in the Sierra Nevada. Greg Lamarre, a scientist at the Smithsonian Tropical Research Institute, reported that insect populations on Panama's Barro Colorado Island appeared to be stable. Dan Janzen, a tropical ecologist who teaches part of the year at the University of Pennsylvania, spoke from his home in northwestern Costa Rica; behind him hung plastic bags filled with what I assumed were specimens. In the sixty years that he'd been studying the region, he said, insect numbers had fallen catastrophically.

"When conservationists speak about tropical forests that have been heavily hunted, they call them 'empty forests,'" he observed. "What we're seeing is an empty forest."

Wagner told me that the heterogeneity of the data coming in had convinced him that insects, rather than suffering from one particular thing, were suffering from everything. "If the stressors are the things we understand—such as lights, pesticides, loss of habitat, climate change, pollution, exotic species, and the industrialization of agriculture—I think that makes perfect sense," he said. The main threat in the American West, he believes, is drought driven by warming.

"Insects are all surface area and no volume," he explained. "So they don't have the capacity to store water. And they're additionally challenged, because their respiratory system differs from ours. We take in air through our mouths and deliver oxygen through our blood. But insect blood doesn't carry oxygen. So they have to have these breathing tubes that penetrate every single cell group in their body, and that compounds their rate of water loss." Insects endemic to dry areas must, of course, have ways of dealing with drought, or they wouldn't exist. Some moths, for example, can wait out a dry spell underground, as pupae, in a state close to suspended animation. But even they have their limits. At a certain point, with no rain, the pupae expire. Wagner said he felt he was in a race against aridification.

"We're going to solve this climate crisis," he told me. "We're going to decarbonize. But it's going to be too late for a lot of the organisms I love."

ON THE SECOND day of the meeting, everyone put on a safety vest and hard hat and trooped over to the museum's newest addition, under construction on Columbus Avenue. The building, which is slated to open later this year, will house the Solomon Family Insectarium, where live cockroaches, beetles, and leaf-cutter ants will be on display. On the walk over, I fell into conversation with one of Wagner's former graduate students, Piotr Naskrecki, an entomologist who works in Gorongosa National Park, in Mozambique.

"Dave has been my hero and my inspiration pretty much my whole adult life," Naskrecki told me. "He has this ability to reignite my belief that what I'm doing matters. One example comes from when I was working on my PhD. I was doing a side project on the phylogeny of hummingbird flower mites, which live in the nostrils of humming-birds. I showed him the results, and I kind of wanted him to tell me, 'Don't waste your time.' But instead he told me, 'This is fantastic!'"

Although it was a Saturday, the insectarium was humming with construction workers when we arrived. Giant metal flowers rose from the floor, and a twenty-foot-tall amber-colored honeycomb, made from some kind of resin, hung from the ceiling. The place was clearly de-signed to inspire joy and wonder in children (and in whoever might be accompanying them). But I could see from the writing on the wall that the messaging was serious. "Insects appear to be declining globally, in ways we are only beginning to understand," a newly painted placard said.

As it happened, the museum was also showing a temporary exhibit on insects, this one consisting entirely of photographs. Each image was an extreme closeup of a specimen from the museum's vast insect collection, with an emphasis on species that are endangered or already gone. During one of the breaks in the meeting, I wandered around the exhibit, which was nearly empty, though the rest of the museum was mobbed. Two boys were fiddling with the dials on a display monitor that allowed viewers to zoom in even further on the closeups. I heard one of the kids say to the other that he wanted to see the bug's "butthole." The boys' parents exchanged knowing glances.

One of the photos showed an hourglass drone fly, blown up to the size of a Doberman. Each lens on the fly's two compound eyes was visible, as were the three ocelli on the top of its head: these "little eyes," it is thought, help the fly orient itself in space. Hourglass drone flies, the label noted, "were once common throughout much of northern North America, but they have nearly disappeared."

A second photo captured the San Joaquin Valley giant flower-loving fly in profile, with its long, pointy proboscis hanging down like a scabbard. The fly, once thought to be extinct, was rediscovered in two spots in the nineteen-nineties; then, in 2006, one of the populations was obliterated by development. Probably there are just a few hundred individuals left, in a single dune east of Bakersfield, California.

Another photo showed a Rocky Mountain locust with its hind legs extended like ski poles. Rocky Mountain locusts were once so numerous that their swarms blocked the sun. A particularly immense swarm, in 1875, was estimated to stretch over almost two hundred thousand square miles and to comprise three trillion individuals. Thirty years later, the locust was extinct. This story, from trillions to none, is a lot like that of the passenger pigeon, which also disappeared in the early twentieth century. No one knows how the locust was eliminated; probably the cause was farming. "As settlers moved into Native lands in the West, they plowed and planted over the insects' nesting areas," the photo's label said.

In total, there were photos of forty insects on display. All of them were of adults. It occurred to me that, once again, larvae were getting short shrift.

WHEN WAGNER STARTED out in entomology, the only way to be sure what species a caterpillar belonged to was to raise it through metamorphosis and see what emerged. Even then, there were no guarantees: many moths and butterflies look a lot alike, apparently even to each other. To guard against mistakes at mating time, lepidoptera have evolved some of the most elaborate genitalia in the animal world, and to guard against mistaken IDs lepidopterists often had to resort to dissecting a specimen's sex organs under a microscope. (A famous story about Vladimir Nabokov, who served as the curator of lepidoptera at Harvard's Museum of Comparative Zoology in the nineteen-forties, has him announcing to a group of visitors, "Excuse me, I must go play with my genitalia.")

These days, professionals have better options than genital dissection. When Wagner wants to know the identity of a caterpillar he's caught, he euthanizes the insect by injecting it with alcohol. Then he cuts off one of its prolegs—these are the stubby appendages located behind caterpillars' six front, or "true," legs—and sends the tissue off to a lab in Ontario. From this tissue, the lab sequences a section of a gene known as cytochrome c oxidase subunit 1, or CO1. All animals possess a copy of CO1—many copies, actually—which is critical to cellular respiration. But each species' version is slightly different from every other's, so the gene can be used as a kind of taxonomic fingerprint. (What's known as DNA bar coding usually involves an analysis of CO1.)

A few months after we returned from Texas, Wagner got word that the results for the samples he'd sent off from our trip were in. These would tell him what species the caterpillars we'd gathered had belonged to and prove that several were new to science, or, perhaps, do the opposite. Wagner spoke of "unwrapping" the results with the eagerness of a kid looking forward to Christmas. He promised to wait to look at them until we could go through them together, over Zoom. Later, he confessed to me that he had peeked at some of the data early.

The first sequence we unwrapped belonged to the "little green thing" I'd found, which Wagner had led me to believe might be "significant." The caterpillar was basically a pest species, related to a corn earworm, and not significant at all. I hadn't given the green thing much thought—in fact, I couldn't quite remember what it looked like. Nevertheless, it was a blow.

Kelsey Wogan's caterpillar, by contrast, turned out to be a real find. It was a genetic match with a very rare, tan-colored moth known only from two specimens at the Smithsonian. (The moth is so rare it has never been formally described.) Wogan was probably the only person ever to have collected one of the moth's rose-colored larvae, and Wagner was the only person ever to have photographed it. "You can't do much better than that," he said.

In among the results were several more hits. The suspected *U. furtiva* caterpillar did, indeed, turn out to be an *Ursia furtiva*. A bright-yellow number that Wagner had collected near the Mexican border turned out to belong to another extremely rare species, *Neoilliberis arizonica;* probably it, too, was the first caterpillar of its kind ever collected. An unremarkable green specimen Wagner had found outside Alpine, the genetic analysis showed, belonged to a species new to science. Another unremarkable caterpillar—a tiny brown inchworm—represented not just a new species but possibly a new genus. There was also a surprise new species Wagner asked me not to reveal much about. This was to prevent other entomologists from trying to describe it before he had a chance to.

"You could say, 'There was one exceptionally interesting discovery that Dave made me embargo,' " Wagner suggested during one of our Zoom sessions.

"The more you know, the more fun this is," he said during another.

Wagner once told me, matter-of-factly, that he had never experienced a period of gloom or depression. Apparently, this included the time he spent going to school with kids who threw rocks at him, as well as the years he has devoted to pondering the "insect apocalypse." During the week I spent with him in Texas, I never saw him get rattled or annoyed, even though one day he sat on (and shattered) his iPhone and another day he had to skip an elaborate expedition he'd planned because a colleague reminded him, a few hours before the deadline, that he hadn't filed a student recommendation. At the conference at the Natural History Museum, he never seemed down, even when the presentations appeared to leave no other option. I wasn't quite sure what to make of this. Was his optimism simply a matter of temperament, or did it have something to do with looking at life through a loupe?

"To a person attuned to smaller creatures," Wagner's former professor Howard Ensign Evans once wrote, "there is no corner of nature not full of excitement, not rich in unsolved problems."

The Earth, Evans added, "is a good place to live."

Originally published in *The New Yorker* as "A Little-Known Planet," March 20, 2023.

KILLING MRS. TIGGY-WINKLE

New Zealand Tries to Rid Itself
of Invasive Species

I N THE DAYS—perhaps weeks—it had spent in the trap, the stoat had lost most of its fur, so it looked as if it had been flayed. Its exposed skin was the deep, dull purple of a bruise, and it was coated in an oily sheen, like a sausage. Stoat traps are often baited with eggs, and this one contained an empty shell. Kevin Adshead, who had set the trap, poked at the stoat with a screwdriver. It writhed and squirmed, as if attempting to rise from the dead. Then it disgorged a column of maggots.

"Look at those teeth," Adshead said, pointing with his screwdriver at the decomposing snout.

Adshead, who is sixty-four, lives about an hour north of Auckland. He and his wife, Gill, own a thirty-five-hundred-acre farm, where for many years they raised cows and sheep. About a decade ago, they decided they'd had enough of farming and left to do volunteer work in the Solomon Islands. When they returned, they began to look at the place differently. They noticed that many of the trees on the property, which should have been producing cascades of red flowers around Christmastime, instead were stripped bare. That was the work of brushtail possums. To save the trees, the Adsheads decided to eliminate the possums, a process that involved dosing them with cyanide.

One thing led to another, and soon the Adsheads were also going after rats. With them, the preferred poison is an anticoagulant that causes internal hemorrhaging. Next came the stoats, or, as Americans would say, short-tailed weasels. To dispatch these, the Adsheads lined their farm with powerful traps, known as DOC 200s, which feature spring-controlled kill bars. DOC 200s are also helpful against ferrets, but the opening is too small for cats, so the Adsheads bought cat traps, which look like rural mailboxes, except that inside, where the letters would go, there's a steel brace that delivers an uppercut to the jaw.

The Adsheads put out about four hundred traps in all, and they check them on a regular rotation. When I visited, on a bright blue day toward the end of the Southern Hemisphere winter, they offered to show me how it was done. They packed a knapsack of supplies, including some eggs and kitty treats, and we set off.

As we tromped along, Kevin explained his trapping philosophy. Some people are fastidious about cleaning their traps of bits of rotted stoat. "But I'm not," he said. "I like the smell in there; it attracts things." Often, he experiments with new techniques; recently he'd learned about a kind of possum bait made from flour, molasses, and cinnamon, and Gill had whipped up a batch, which was now in the knapsack. For cats, he'd found that the best bait was Wiener schnitzel.

"I slice it thin and I tie it over the trigger," he told me. "And what happens with that is it starts to dry out and they still go for it."

I'd come to watch the Adsheads poke at decaying stoats because they are nature lovers. So are most New Zealanders. Indeed, on a per capita basis, New Zealand may be the most nature-loving nation on the planet. With a population of just four and a half million, the country has some four thousand conservation groups. But theirs is, to borrow E. O. Wilson's term, a bloody, bloody biophilia. The sort of amateur naturalist who in Oregon or Oklahoma might track butterflies or band birds will, in Otorohanga, poison possums and crush the heads of hedgehogs. As the coordinator of one volunteer group put it to me, "We always say that, for us, conservation is all about killing things."

The reasons for this are in one sense complicated—the result of a peculiar set of geological and historical accidents—and in another quite simple. In New Zealand, anything with fur and beady little eyes is an invader, brought to the country by people—either Maori or European settlers. The invaders are eating their way through the native fauna, producing what is, even in an age of generalized extinction, a major crisis. So dire has the situation become that schoolchildren are regularly enlisted as little exterminators. (A recent blog post aimed at hardening hearts against cute little fuzzy things ran under the headline "Mrs. Tiggy-Winkle, Serial Killer.")

Not long ago, New Zealand's most prominent scientist issued an emotional appeal to his countrymen to wipe out all mammalian predators, a project that would entail eliminating hundreds of millions, maybe billions, of marsupials, mustelids, and rodents. To pursue this goal—perhaps visionary, perhaps quixotic—a new conservation group was formed this past fall. The logo of the group, Predator Free New Zealand, shows a kiwi with a surprised expression standing on the body of a dead rat.

NEW ZEALAND CAN be thought of as a country or as an archipelago or as a small continent. It consists of two major islands—the North Island and the South Island, which together are often referred to as the mainland—and hundreds of minor ones. It's a long way from anywhere, and it's been that way for a very long while. The last time New Zealand was within swimming distance of another large landmass was not long after it broke free from Australia, eighty million years ago. The two countries are now separated by the twelve-hundred-mile-wide Tasman Sea. New Zealand is separated from Antarctica by more than fifteen hundred miles and from South America by five thousand miles of the Pacific.

As the author David Quammen has observed, "Isolation is the flywheel of evolution." In New Zealand, the wheel has spun in both directions. The country is home to several lineages that seem impossibly outdated. Its frogs, for example, never developed eardrums, but, as if in compensation, possess an extra vertebra. Unlike frogs elsewhere, which absorb the impact of a jump with their front legs, New Zealand frogs, when they hop, come down in a sort of belly flop. (As a recent scientific paper put it, this "saltational" pattern shows that "frogs evolved jumping before they perfected landing.") Another "Lost World" hold-over is the tuatara, a creature that looks like a lizard but is, in fact, the sole survivor of an entirely separate order—the sphenodonts—which thrived in the early Mesozoic. The sphenodonts were thought to have vanished with the dinosaurs, and the discovery that a single species had somehow managed to persist has been described as just as sur-prising to scientists as the capture of a live *Tyrannosaurus rex* would have been.

At the same time, New Zealand has produced some of nature's most outlandish innovations. Except for a few species of bats, the country has no native mammals. Why this is the case is unclear, but it seems to have given other groups more room to experiment. Weta, which re-semble giant crickets, are some of the largest insects in the world; they scurry around eating seeds and smaller invertebrates, playing the part that mice do almost everywhere else. *Powelliphanta* are snails that seem to think they're wrens; each year, they lay a clutch of hard-shelled eggs. *Powelliphanta,* too, are unusually big—the largest measure more than three and a half inches across—and, in contrast to most other snails, they're carnivores, and hunt down earthworms, which they slurp up like spaghetti.

New Zealand's iconic kiwi is such an odd bird that it is sometimes referred to as an honorary mammal. When it was first described to En-glish naturalists, in 1813, they thought it was a hoax. Kiwi are covered in long, shaggy feathers that look like hair, and their extended, tapered beaks have nostrils on the end. They are around the size of chickens but lay eggs that are ten times as large, and it usually falls to the male, Horton-like, to hatch them.

New Zealand's biggest oddballs were the moa, which, in a feathers-for-fur sort of way, stood in for elephants and giraffes. The largest of them, the South Island giant moa, weighed five hundred pounds, and with its neck outstretched could reach a height of twelve feet. Moa fed on New Zealand's native plants, which, since there were no mammal-ian browsers, developed a novel set of defenses. For instance, some New Zealand plants have thin, tough leaves when they are young, but when they mature—and grow taller than a moa could chomp on—

they put out leaves that are wider and less leathery. The Australian naturalist Tim Flannery has described New Zealand's avian-dominated landscape as a "completely different experiment in evolution." It shows, he has written, "what the world might have looked like if mammals as well as dinosaurs had become extinct 65 million years ago, leaving the birds to inherit the globe." Jared Diamond once described the country's native fauna as the nearest we're likely to come to "life on another planet."

WITH ABOUT HALF the population of New Jersey, New Zealand is the sort of place where everyone seems to know everyone else. One day, without quite understanding how the connection had been made for me, I found myself in a helicopter with Nick Smith, who was then the country's minister for conservation.

Smith, who is forty-nine, has a ruddy face and straw-colored hair. He is a member of the country's center-right National Party, and calls himself a "Bluegreen." (Blue is the National Party's color.) He got interested in politics back in the nineteen-seventies, when, as an exchange student in Delaware, he met Joe Biden. We set off from Smith's district office, in the city of Nelson, on the northern tip of the South Island, and drove out to the helicopter pad in his electric car.

"When the first settlers came here, they tried to create another England," Smith told me. "We were Little Britain. The comment about us was, we were more British than the British. And, as part of the maturing of New Zealand, there's the question 'What do you connect your nationhood to?' You know, for America it's very much the flag, the Constitution, those sorts of things. The connection with species that are unique to New Zealand is increasingly part of our national identity. It's what we are as New Zealanders, and I make no bones of the fact that the government is keen to encourage that. You need some things for a country to hold together."

He went on, "I say to people, 'If you want your grandkids to see kiwi only in sanctuaries, well, that's where we're headed.' And that's why we need to use pretty aggressive tools to try to turn this around."

My visit happened to coincide with the application of one of these aggressive tools. The country's Department of Conservation was conducting a massive aerial drop of a toxin known as 1080. (The key ingredient in 1080, sodium fluoroacetate, interferes with energy production on a cellular level, inflicting what amounts to a heart attack.) New Zealand, which has roughly one-tenth of 1 percent of the world's land, uses 80 percent of its 1080. This year's drop—the department was planning to spread 1080 over nearly two million acres—had been prompted by an unusually warm winter, which had produced an exceptionally large supply of beech seed, which in turn had produced an explosion in the number of rats and stoats. When the beech seed ran out, the huge cohort of predators was expected to turn its attention to the native fauna. Smith had approved the 1080 operation, which had been dubbed Battle for Our Birds, but the timing of it troubled him; owing to the exigencies of rat biology, the drop had to take place right around the time of a national election. "If you ask the cynical politics of it, people don't like poisons but they like rats even less," he told me. "And so I've been doing a few quite deliberate photo opportunities with buckets of rats."

On this particular day, Smith was attending a more cheerful sort of photo op—one with live animals. When we arrived at the helicopter pad, three other people were already there, all associated with a privately funded effort to restore one of the country's most popular national parks, named for Abel Tasman. (In 1642, Tasman, a Dutch explorer, was the first European to reach New Zealand—though he didn't quite reach it, as four of his sailors were killed by Maori before they could land.) Smith and I got into the helicopter next to the pilot, the other three climbed into the back, and we took off. We flew over Tasman Bay and then over the park, which was studded with ghostly white trees.

"We like to see all those dead trees," Devon McLean, the director of the restoration project, announced cheerfully into his headset. He explained that the trees were invasive pines, known in New Zealand as "wilding conifers." I had a brief vision of scrawny seedlings rampaging

through the forest. Each dead tree, McLean said, had been individually sprayed with herbicide. He was also happy to report that the park
had recently been doused with 1080.

After about half an hour, we landed on a small island named Adele,
where we were greeted by a large sign: "Have You Checked for Rats,
Mice and Seeds?" A few years ago, after an intensive campaign of poi-
soning and trapping, Adele was declared "pest-free." The arrival of a
single pregnant rat could undo all that work; hence the hortatory sign-
age. In another helicopter, the Conservation Department was going to
deliver two or three dozen representatives of one of New Zealand's
rarest species, the South Island saddleback. The birds would be released onto Adele, where, it was hoped, in the absence of rats, they
would multiply.

More people began to arrive by boat—reporters, representatives of
several regional conservation groups, members of the local Maori *iwi,*
or tribe. By this point, it was drizzling, but there was a festive mood on
the beach, as if everyone were waiting for a celebrity. "Hardly any New
Zealanders have ever seen a saddleback," a woman whispered to me.

In anticipation of the birds' appearance, speeches were offered in
Maori and English. "It has taken us Pakeha New Zealanders a little
while to gain an appreciation of what is special here and to really be
committed to its protection and survival," Smith said, using the
Maori word for "European." "It's kind of scary to think of South Island saddlebacks—there are only about six hundred that exist on the
planet."

Finally, the birds arrived, in a helicopter that had been loaned for
the occasion by a wealthy businessman. Three crates were unloaded
onto the beach, and Smith and a pair of local dignitaries were given the
honor of opening them. South Island saddlebacks are glossy black,
with patches of rust-colored feathers around their middles and little
orange wattles that make them look as though they're smiling. They
are another example of an ancient lineage that persists in New Zealand, and they have no close relatives anywhere else in the world. The
birds hopped out of their crates, flew into the bush, and were gone.

• • •

THERE ARE TWO quick ways to tell a Norway rat from a ship rat. One is to look at the ears. Ship rats have large ears that stick out from their heads; Norway rats' ears are shorter and less fleshy. The other is to look at their tails. Again, Norway rats' are shorter. With a ship rat, if you take its tail and fold it over its body—here it obviously helps if the rat is dead—it will extend beyond its nose.

These and many other facts about rats I learned from James Russell, an ecologist at the University of Auckland. Russell's office is filled with vials containing rat body parts in various stages of decomposition, and he also keeps a couple of dead rats at home in his freezer. Wherever he goes, Russell asks people to send him the tails of rats that they have trapped, and often they oblige. Russell then has the rats' DNA sequenced. Eventually, he hopes to be able to tell how all of New Zealand's rat populations are related.

"I would be inclined to say rats are our biggest problem," Russell told me. "But I have colleagues who spend their career on stoats, and colleagues who spend their career on cats. And they open all their talks with 'Stoats are the biggest problem' or 'Cats are the biggest problem.'"

Russell, who is thirty-five, is a slight man with tousled brown hair and a cheerful, let's-get-on-with-it manner that I eventually came to see as very New Zealand. I ended up spending a lot of time with him, because he volunteered to guide me along some of the country's windier back roads.

"New Zealand was the last large landmass on earth to be colonized," he told me one day, as we zipped along through the midsection of the North Island. "And so what we saw was the tragedy of human history playing out over a short amount of time. We're only ten years behind a lot of these things; that's as compared to countries where you're hundreds or thousands of years behind the catastrophe."

New Zealand's original settlers were the Maori, Polynesians who pitched up around the year 1300, probably from somewhere near the Society Islands. By that point, people had already been living in Australia for some fifty thousand years. They'd been in continental North America for at least ten thousand years, and in Hawaii, which is even more remote than New Zealand, for more than five hundred years. In each case, it's now known, the arrival of humans precipitated a wave of extinctions; it's just that, as Russell points out, these "tragedies" were not recorded by the people who produced them.

When the Maori showed up, there were nine species of moa in New Zealand, and it was also home to the world's largest eagle—the Haast's eagle—which preyed on them. Within a century or two, the Maori had hunted down all of the moa, and the Haast's eagles, too, were gone. A Maori saying—*Ka ngaro i te ngaro a te moa*—translates as "Lost like the moa is lost."

In their ships, the Maori also brought with them Pacific rats, or kiore. These were New Zealand's first introduced mammals (unless you count the people who brought them). The Maori intended to eat the kiore, but the rats multiplied and spread far faster than they could be consumed, along the way feasting on weta, young tuatara, and the eggs of ground-nesting birds. In what, in evolutionary terms, amounted to no time at all, several species of New Zealand's native ducks, a couple of flightless rails, and two species of flightless geese were gone.

The arrival of British settlers, in the middle of the nineteenth century, brought in more—many more—new invaders. Some of them, like the Norway rat and the ship rat, were stowaways. Others were introduced deliberately, in an effort to make New Zealand feel more like home. The "importation of those animals and birds, not native to New Zealand," an 1861 act of the colonial parliament declared, would both "contribute to the pleasure and profit of the inhabitants" and help maintain "associations with the Old Country." What were known as "acclimatization societies" sprang up in every region. Among the many creatures the societies tried to "acclimatize" were red deer, fallow deer, white-tailed deer, sika deer, tahr, chamois, moose, elk, hedgehogs, wallabies, turkeys, pheasants, partridges, quail, mallards, house sparrows, blackbirds, brown trout, Atlantic salmon, herring, whitefish, and carp. Brushtail possums were specially imported from Australia, in an attempt to start a fur industry.

Not all the new arrivals took; others took all too well. By the perverse logic of such affairs, some of the most disastrous introductions were made in an effort to control previous disastrous introductions. Within a few decades of their importation, European rabbits had overrun the countryside, and in 1876 an act was passed to "provide for the Destruction of Rabbits in New Zealand." The act had no perceptible impact, so stoats and ferrets were released into the bush in the hope that they would be more effective.

"Our forebears tried most experiments that could possibly be conceived and some that would be difficult for anyone with any knowledge of ecology to seriously contemplate," Robert McDowall, a New Zealand naturalist, has written in a history of introduction efforts. The combined effect of all these "experiments," particularly the introduction of predators, like stoats, has been ongoing devastation. Roughly a quarter of New Zealand's native bird species are now extinct, and many of those that remain are just barely hanging on. If current trends continue, it is predicted that within a generation or two the land of Kiwis will be without kiwi.

"The defence of isolation for remote islands has no fallback position," John McLennan, a New Zealand ecologist, has written. "It is all or nothing, akin to virginity, with no intermediate state."

SIROCCO, A SEXUALLY dysfunctional kakapo, normally lives alone on his own private island. On occasion, though, he is brought, with great fanfare, to the mainland, and this is where James Russell and I set out to meet him, at a special mountaintop reserve ringed by a twenty-nine-mile fence. The fence is seven feet high and made of steel mesh with openings so narrow an adult can't stick a pinkie through. At the base of the fence, an eighteen-inch apron prevents rats from tunneling under; on top, an outwardly curved metal lip stops possums and feral cats from clambering over. To get inside, human visitors have to pass through two sets of gates, an arrangement that made me think of a maximum-security prison turned inside out.

Kakapo—the name comes from the Maori, meaning "parrot of the night"—are nocturnal, so all audiences with Sirocco take place after dark. Russell and I joined a group of about twenty other visitors, who had paid forty dollars apiece to get a peek at the bird. Sirocco was hopping around in a dimly lit plate-glass enclosure. He was large—about the size of an osprey—with bright green-and-brown feathers and a bulbous, vaguely comical beak. He gazed through the glass at us and gave a sharp cry.

"He's very intense, isn't he?" the woman next to me said.

Alone among parrots, kakapo are flightless, and, alone among flightless birds, they're what's known as lek breeders. During mating season, a male will hollow out a little amphitheater for himself, puff up his chest, and let out a "boom" that sounds like a foghorn. Kakapo, which can live to eighty, breed only irregularly, in years when their favorite foods are in good supply.

Kakapo once were everywhere in New Zealand. In the late nineteenth century, they were still plentiful in rugged areas; Charlie Douglas, an explorer who climbed some of the steepest mountains of the South Island, described them standing "in dozens round the camp, screeching and yelling like a lot of demons." But then their numbers crashed. By the nineteen-seventies, there was just one small population remaining, and it was threatened by feral cats. In the nineteen-eighties, every individual that could be caught was captured and "translocated." Today, there are a hundred and twenty-six kakapo left, and all of them, save Sirocco, live on three remote, predator-free islands—Little Barrier, Codfish, and Anchor.

Sirocco's chaperone, a Department of Conservation ranger named Alisha Sherriff, had brought along a little metal container, which she passed among the visitors. Inside was half a cup's worth of Sirocco's shit.

"Have a good sniff," she suggested.

"It's earthy!" one woman exclaimed.

"I think it smells smoky, with notes of honey," Sherriff said. When the container came to me, I couldn't detect any honey, but the bouquet did strike me as earthy, with hints of newly mown hay. Sherriff had also brought along a ziplock bag with some of Sirocco's feathers. These, too, had a strong, sweetish scent. New Zealand birds tend to smell, which was not a problem when the islands' top predators were avian, since birds hunt by sight. But, as mammals hunt with their noses, it's become yet another liability. (A few years ago, a Christchurch biologist was awarded a four-hundred-thousand-dollar grant to investigate the possibility of developing some sort of "deodorant" for ground-nesting birds.)

Sherriff explained that Sirocco had been born on Codfish Island in 1997, but as a chick he had come down with a respiratory infection and so had been removed from his mother and raised in isolation. By the time he was well enough to rejoin the other kakapo, he'd decided he preferred people. During breeding season, he'd try to mate with the rangers on Codfish while they were walking to the outhouse. A special barrier was built to try to prevent the encounters, but Sirocco turned out to be too determined.

"When you've got a three-kilo bird crashing through the bush in the middle of the night, there's quite a high risk of people lashing out unintentionally," Sherriff said. It had been decided that, for Sirocco's own safety, he'd have to be moved. He now lives hundreds of miles from Codfish, on an island whose name the Department of Conservation won't disclose. For the sake of genetic diversity, Sirocco's semen had been collected, by means of what Sherriff described as a "delicate massage," but his sperm count had proved too low to attempt artificial insemination.

After about half an hour with Sirocco, we were hustled out to make room for the next tour group. Russell and I passed back through the gates and drove down to a restaurant at the base of the mountain, where we'd arranged to have dinner with Matt Cook, the reserve's natural-heritage manager. Cook told us that it had taken teams of exterminators three years to eliminate mammals from inside the fence and that they'd never managed to finish off the mice. The entire perimeter was wired so that when, say, a section of fence was hit by a falling branch, a call automatically went out to a maintenance crew.

"This always happens at three a.m. on a Saturday morning," he said. If the fence was breached, Cook reckoned, the crew had about ninety minutes to repair it before rats would find the opening and sweep back in.

. . .

TODAY, INVASIVE SPECIES are everywhere. No matter where you are reading this, almost certainly you are surrounded by them. In the northeastern United States, common invasive plants include burdock, garlic mustard, purple loosestrife, and multiflora rose; invasive birds include starlings, rock pigeons, and house sparrows; and invasive insects include Japanese beetles, gypsy moths, and hemlock woolly adelgid. Texas has more than eight hundred non-native plant species, California at least a thousand. Even as New Zealand has been invaded by non-native species, its native species have invaded elsewhere. *Potamopyrgus antipodarum,* or, as it is more commonly known, the New Zealand mud snail, is a tiny aquatic snail about the size of a grain of rice. It can now be found in Europe, Asia, Australia, the Middle East, and the American West, and it has proved such a successful transplant that in some parts of the world it reaches densities of half a million snails per square yard. In an irony that perhaps only Kiwis can appreciate, a recent study in Utah found that mud snails were threatening populations of rainbow trout, a fish imported at great expense to New Zealand in the eighteen-eighties and now considered an invader there.

The project of reshuffling the world's flora and fauna, which began slowly with the spread of species like the Pacific rat and sped up thanks to the efforts of acclimatization societies, has now, with global trade and travel, accelerated to the point that, on any given day, something like ten thousand species are being moved around just in the ballast water of supertankers. Such is the scale of this remix that biologists have compared it to reassembling the continents. Two hundred million years ago, all of the world's landmasses were squished together into a single giant supercontinent, Pangaea. We are, it's been suggested, creating a "new Pangaea."

One response is simply to accept this as the planet's destiny. Yes, the invaders will, inevitably, choke out some local species, and there will be losses, especially on islands, where, unfortunately, much of the world's diversity resides. But people aren't going to stop shipping goods and they aren't going to stop traveling; therefore, we're just going to have to learn to live in a Pangaea of our own making. Meanwhile, who even knows at this point what's native and what's not? Many species that people think of as a natural part of the landscape are really introductions that occurred before recent memory. For instance, the ring-necked pheasant, the official state bird of South Dakota, is an import from China.

"It is impractical to try to restore ecosystems to some 'rightful' historical state," a group of American researchers wrote a few years ago, in *Nature*. "We must embrace the fact of 'novel ecosystems' and incorporate many alien species into management plans, rather than try to achieve the often-impossible goal of eradicating them."

New Zealanders are nothing if not practical. They like to describe the national mindset as "the No. 8 wire mentality"; for much of the country's history, No. 8 wire was used to fix livestock fences and just about everything else. Nevertheless, Kiwis refuse to "embrace" novel ecosystems. In the past few decades, they have cleared mammalian predators from a hundred and seventeen offshore islands. The earliest efforts involved tiny specks, like Mokopuna, or, as it is also known, Leper Island, which is about the size of Gramercy Park. But, more recently, they've successfully de-ratted much larger islands, like Campbell, which is the size of Nantucket. With its predator-free islands and its fenced-in reserves and its massive poison drops from the air, New Zealand has managed to bring back from the very edge of oblivion several fantastic birds, including the kakapo, the South Island saddleback, the Campbell Island teal, and the black robin. At its lowest point, the black robin was down to just five individuals, only one of which—a bird named Old Blue—was a fertile female. (When Old Blue died, her passing was announced in Parliament.)

Meanwhile, by tackling larger and larger areas, New Zealanders have expanded the boundaries of what seems possible, and they increasingly find their skills in demand. When, for example, Australia decided to try to get rid of invasive rodents on Macquarie Island, roughly halfway between Tasmania and Antarctica, it hired a New Zealander to lead the effort, and when the U.S. National Park Service decided to get rid of pigs on Santa Cruz Island, off the coast of Southern California, it hired Kiwis to shoot them. The largest rat-eradication effort ever attempted is now in progress on South Georgia Island, a British territory in the South Atlantic with an area of nearly a million acres. A New Zealand helicopter pilot was brought in to fly the bait-dropping missions. One day, when I was driving around with James Russell, he got an email from Brazil: the government wanted to hire him to help it get rid of rats on the Fernando de Noronha archipelago, off Recife. David Bellamy, a British environmentalist and TV personality, has observed that New Zealand is the only country in the world that has succeeded in turning pest eradication into an export industry.

THE IDEA OF ridding all of New Zealand of its mammalian predators was proposed by Paul Callaghan, a world-renowned physicist, in a speech delivered in Wellington in February 2012. In scientific circles, Callaghan was celebrated for his work on nuclear magnetic resonance; to Kiwis he was probably best known for having recently been named New Zealander of the Year. At the time he gave the speech, Callaghan was dying of cancer, and everyone who heard it realized that it was one of the last he would deliver. (He died the following month.)

"Let's get rid of the lot," Callaghan said. "Let's get rid of all the predators—all the damned mustelids, all the rats, all the possums— from the mainland."

"It's crazy," he continued, referring to his own proposal. "But," he went on, "I think it might be worth a shot. I think it's our great challenge." Callaghan compared the project to the moon landing. It could be, he said, "New Zealand's Apollo program."

I listened to Callaghan's speech, via YouTube, on one of my last evenings in New Zealand. It seemed to me that what he was proposing was more like New Zealand's Manhattan Project than its Apollo program, though I could see why he hadn't framed it that way.

New Zealand's North Island is roughly forty-four thousand square miles. That means that it's nearly a thousand times bigger than the largest offshore island from which predators have, at this point, been eradicated. James Russell serves as an adviser to Predator Free New

Zealand, the group that was formed to pursue Callaghan's vision. I asked him about the feasibility of scaling up by such an enormous amount. In response, he showed me a graph: the size of the islands from which predators have been successfully removed has been increasing by roughly an order of magnitude each decade.

"It is a daunting scale that we're talking about," Russell told me. "But, then, you see the rate at which we have scaled up."

"Some people think scientists should just be objective," he went on. "They sit in the lab, they report their results, and that's it. But you can't separate your private life from your work life. So I do this science and then I go home and think, Wouldn't it be great if New Zealand had birds everywhere and we didn't have to worry about rats? And so that's the world I imagine."

Listening to Callaghan on YouTube also reminded me of a point that Nick Smith had made the day of the saddleback release: in New Zealand, killing small mammals brings people together. During my travels around the country, I found that extermination, weird as it may sound, really is a grassroots affair. I met people like the Adsheads, who had decided to clear their own land, and also people like Annalily van den Broeke, who every few weeks goes out to reset traps in a park near her home, in the suburbs of Auckland. In Wellington, I met a man named Kelvin Hastie, who works for a 3-D mapping company. He had divided his neighborhood into a grid and was organizing the community to get a rat trap into every hundred-square-meter block.

"Most of the neighbors are pretty into it," he told me.

Just about everyone I spoke to, including Hastie, expressed excitement about the latest breakthrough in extermination technology, a device designed by a company called Goodnature. And so, on my last day in the country, I went to visit the company's offices. They were situated on a drab stretch of anonymous buildings, not far from one of Wellington's best surfing beaches. When I arrived, I noticed a dead stoat in a plastic bag by the door. Apparently, it had been killed quite recently, because it was still very much intact.

Robbie van Dam, one of the company's founders, showed me around. In the first room, several young people were working at standing desks, on MacBooks propped up on cardboard boxes. A small kitchen was stocked with candy and half-empty bottles of wine. In a back room, bins of plastic parts were being assembled into an *L*-shaped machine that resembled a portable hair dryer. Van Dam pulled out one of the finished products, known as the A24. At one end, there was a CO_2 cannister of the sort used in bicycle pumps. Van Dam uncapped the other end and pulled out a plastic tube. It was filled with brown goo. In the crook of the *L* was a hole, and in the hole there was a wire. Van Dam gingerly touched the wire, and a piston came flying out.

The A24 is designed to be screwed to a tree. The idea is that a rat, smelling the goo, which is mostly ground nuts, will stick its head into the hole, trip the wire, and be killed instantly. The rat then falls to the ground, and the device—this is the beauty part—automatically resets itself. No need to fish out rotten eggs or decaying flesh. Each CO_2 cannister is supposed to be good for two dozen rats or, alternatively, stoats—hence the name. (For stoats, there's different bait, made of preserved rabbit.) Van Dam also showed me a slightly larger machine, the A12, designed for possums.

"The humaneness problem was probably the hardest part," he told me. In the case of possums, it had turned out that a blow to the head wasn't enough to bring about quick death. For that reason, the A12 is designed to fill the animal's skull with carbon dioxide and emulsify its brain. Goodnature also sells cannisters of possum bait, laced with cinnamon. I picked up a tube. The label said, "12 out of 12 possums choose this as their final meal."

Even taking the long view—the very long view—the threat posed to New Zealand's fauna by invasive species is extraordinary. It may be unprecedented in the eighty million years that New Zealand has existed. But we live in an age of unprecedented crises. We're all aware of them, and mostly we just feel paralyzed, incapable of responding. New Zealanders aren't just wringing their hands, nor are they telling themselves consolatory tales about the birth of "novel ecosystems." They're dividing their neighborhoods into grids and building better possum traps—ones that deliver CO_2 directly to the brain.

A couple of miles from Goodnature's headquarters is a rocky beach where little blue penguins sometimes nest. The beach is infested with rats, which can attack the nests, so Goodnature has installed some A24s along it, and van Dam took me down to have a look. It was a beautiful windy day, and the surf was high. Under the first A24, which was attached to a gnarly bush, there was nothing. Often, van Dam explained, cats or other rats drag off dead animals that have dropped from the A24, so it's not always possible to know what, if anything, has been accomplished. This had proved to be something of a sales problem.

"If people didn't find something dead under there, they were really disappointed," he said. Goodnature now offers a digital counter that attaches to the A24 and records how many times the piston has been released.

Under the second A24, there was a dead mouse; under the third and fourth, nothing. Under the fifth were two ship rats, one freshly killed and the other just a clump of matted fur with a very long tail.

Originally published in *The New Yorker* as
"The Big Kill," December 22 and 29, 2014.

Update: Since I visited New Zealand, a great deal of progress has been made toward eradicating invasive predators. Still, most of the mainland—which is to say tens of thousands of square miles—remains to be cleared. In 2025, work is supposed to begin on eradicating rats and stoats from Raki-ura, also known as Stewart Island, which is six hundred and fifty square miles. If the project succeeds, Rakiura will become New Zealand's largest predator-free zone. Eliminating all of New Zealand's introduced predators by 2050 is now the country's official goal.

STUNG

Where Have All the Bees Gone?

NOT LONG AGO, I found myself sitting at the edge of a field with a bear and thirty or forty thousand very angry bees. The bear was there because of the bees. Why I was there was a question I found myself unable to answer precisely.

In a roundabout sort of way, the encounter had been set in motion several months earlier, late in February, when the *The New York Times* ran a story about a new ailment afflicting honeybees. It had been given a name—colony-collapse disorder—but no one had any idea what was causing it; beekeepers would open their hives only to discover that they were suddenly mysteriously empty. According to the article, some keepers had lost 70 percent of their colonies, and these losses, in turn, were likely to reduce the yields of crops ranging from kiwis to avocados. All this information struck me as disturbing, and therefore interesting. I thought that at some point I might want to write about it myself, and so I began to read up on bees.

The literature of apiculture is vast and seductive; I learned one amazing thing after another. Honeybees are the only animals besides humans known to have representational language: they convey to one another the location of food by dancing. When the queen lays an egg, she is able to choose its sex. Males, known as drones, perform no useful function except to mate. They are loutish and filthy, and the workers— sterile females—tolerate their presence for a few months a year, then systematically murder them. A single pound of clover honey represents the distilled nectar of some 8.7 million flowers. In a week, a productive hive can add seventy pounds of honey to its stores. Pretty soon, I had moved on to beekeeping manuals. I learned about different "races" of honeybees, each with its own "dialect" and disposition: Italians, which are golden and laid-back but can have trouble producing enough honey for winter; Carniolans, which are darker and hardier but prone to swarm; and Russians, which build up slowly but are the hardiest of all. I also learned about honeybee diseases: varroa mites, tiny parasites that attach themselves to bees and feed on their blood; tracheal mites, even tinier parasites, which attack bees' breathing tubes; American foulbrood, bacteria that turn bee larvae into stringy goo; and sac brood, a virus that leaves larvae swimming in bubbles of muck. Finally, and, I suppose, predictably, I began leafing through beekeeping catalogs, weighing the advantages of wooden frames versus plastic ones and full-body "English-style" bee suits versus simpler (and cheaper) veils. By the time I ordered my hive, the initial reason for having one—to learn about colony-collapse disorder—had dissipated. The disease (or whatever it was) hadn't turned up in the region where I live, which is western Massachusetts. But by that point I wasn't sure whether I was writing the story to keep bees or keeping bees to write the story.

BEES, WHICH ARE descended from predatory wasps, turned from eating other insects to feeding on flowers some hundred million years ago. Not coincidentally, this was shortly after flowers first appeared. Since then, it's been one long evolutionary tango. Some bees have evolved to feed on the nectar of a single type of flower: for example, *Andrena florea,* a small European bee, relies exclusively on the delicate white blossoms of bryony plants. Conversely, some flowers, like the showy, fragrant orchid *Stanhopea embreei,* native to Ecuador, are pollinated only by a single species of bee, in this case *Eulaema bomboides.* Worldwide, nearly twenty thousand species of bees have been identified. Out of these, only perhaps two dozen have been successfully raised by humans, and only one—*Apis mellifera,* commonly known as the western honeybee—accounts for nearly all the bees maintained by beekeepers in Europe and North America.

A. mellifera is a floral generalist—the technical term is *polylectic*—meaning that it will feed on just about anything that is blooming. This trait makes honeybees essential to modern agriculture, which has itself evolved to depend on their services. In a five-hundred-acre apple orchard, for example, there simply aren't enough indigenous pollinators to produce a commercial crop: either the yield will be too low or the fruit will be small or stunted. (An apple has ten ovules, each of which can produce a seed; unless at least six are pollinated, the apple will be misshapen.) Apple growers, therefore, bring additional pollinators into their fields, and honeybees are the only ones that can be delivered in sufficient numbers. Other commercial crops that have come to rely on honeybees include blueberries, cranberries, cherries, cucumbers, watermelons, cantaloupes, and pumpkins.

Almonds, in particular, have extremely high pollination requirements—nearly all the flowers in an orchard must be cross-pollinated to produce a commercial crop—and so California's increasingly large (and lucrative) almond industry is almost entirely honeybee dependent; it is estimated that to service the state's two billion dollars' worth of almonds next year will require nearly 1.5 million hives, or roughly two-thirds of all the colonies that existed in this country before colony-collapse disorder. (The price of renting a hive during the almond bloom, which starts in late February, rose from fifty-five dollars three years ago to a hundred and thirty-five dollars this year, and next year will likely reach a hundred and seventy dollars.) Five years from now, as more acreage goes into production, it is expected that almonds will require 2.1 million colonies, or nearly all the hives that are currently being kept, both by commercial beekeepers and by hobbyists.

Typically, commercial beekeepers ship their hives by flatbed truck; the hives are stacked on pallets, then unloaded with a forklift. The process is efficient—two men can easily move ten million bees into an orchard in a single day—and also profitable, or at least more profitable than selling honey to a world drenched in corn syrup. But it is hard on the bees. One keeper told me that every time he loads up his hives he expects to lose 10 percent of his queens simply as a result of the jostling. Insecticides are also a problem; even assuming that farmers are careful to avoid spraying when bees are in their fields—something that beekeepers complain is not always the case—there are residues. Finally, the mass movement of honeybees spreads parasites and disease. (Truck 1.5 million hives in to pollinate almonds, and some sixty billion bees will be buzzing around the same trees.) The blood-sucking varroa mite was first reported here in 1987; within a few years, practically every managed colony in the United States had been infected. Since the early eighties, the number of hives has dropped by almost half. Wild honeybees, meanwhile, which were once common across the country, have nearly disappeared.

THE FIRST PERSON to notice that there was something seriously wrong with his bees—or maybe just the first person to admit it—was David Hackenberg. Hackenberg, who is fifty-eight, is the owner, founder, and chief source of labor for Hackenberg Apiaries, which is based in Lewisburg, Pennsylvania. He has a weathered face, grayish-brown eyes, and legs that seem to take up three-quarters of his body. He has been keeping honeybees for forty-five years.

"I started with one hive for a Vo-Ag project," Hackenberg told me not long ago. We were standing in a field somewhere near Lake Ontario, and a few yards away his oldest son, Davey, was collecting boxes of honey. There were so many bees in the air that it seemed to be vibrating. "I thought there was money in it. And there is. I keep putting it there." By the time Hackenberg graduated from high school, he had a hundred and fifty hives. That figure kept on growing until it reached nearly three thousand.

Last year, as usual, Hackenberg spent the spring ferrying his hives from Pennsylvania, where the bees pollinated apples, to Maine, where they worked low-bush blueberries, to upstate New York, where they fed on clover, and finally back to Pennsylvania, where they pollinated pumpkins. In October, Hackenberg and his son trucked the bees down to Florida for the winter. They left four hundred hives on a lot south of Tampa, so that the bees could feed off an invasive weed, Brazilian pepper, that was blooming nearby. In mid-November, they returned to pick up the hives, because the owner of the lot—a man who rents out carnival rides—needed it to store equipment. At that point, they did what they always do—put on their protective gear and lit a smoker. (For reasons that are not entirely clear, bees respond to the smell of burning wood or straw by becoming more docile, so beekeepers usually smoke hives before handling them.)

"After I smoked about five pallets, I realized I'm not smoking anything," Hackenberg recalled. "I started jerking covers off, and the hives are empty." Increasingly frantic, he began pulling the frames out of the hives. The more he saw, the weirder the situation looked. The frames all had honey in them, indicating that there had been plenty of food. They were filled with young larvae, or brood, meaning that the bees, usually fiercely maternal, had abandoned their young. There were no signs of moths or other pests that normally invade sick colonies. And Hackenberg couldn't find any dead bees.

"I got down on my hands and knees looking," he told me. "They weren't there. It's like somebody swept the boxes out. There were just no bees." Every time he came across a dead hive—what beekeepers refer to as a deadout—he flipped it on its side. By the time he had gone through the four hundred hives on the lot, only forty were still standing. Hackenberg had shipped twenty-nine hundred hives to Florida. When he finally went through all of them, he found that two thousand had been wiped out.

Hackenberg likes to talk. (Davey told me that one month this spring his father had a cell phone bill for fifty-three hundred minutes.) He began calling around—to fellow beekeepers, to officials at the U.S. Department of Agriculture, to entomologists he knew at Penn State. He told them that some strange new ailment was killing his bees; they told him he had probably just screwed up. "Them mites'll get you," one of Hackenberg's closest friends said. But Hackenberg persisted. Within a week, other beekeepers—people he didn't even know—began calling him to tell him that their bees, too, had disappeared. "I became an expert all of a sudden on something I don't know anything about," Hackenberg said.

All sorts of theories were soon proposed. The mysterious ailment was a new disease, or it was a response to drought, or to stress, or to toxins. According to one widely reported hypothesis, cell phone transmissions were disrupting the bees' navigational abilities. (Few experts took the cell phone conjecture seriously; as one scientist said to me, "If that were the case, Dave Hackenberg's hives would have been dead a long time ago.")

For his part, Hackenberg decided that the culprit was a new class of insecticides, called neonicotinoids. Neonicotinoids are neurotoxins that, as the name implies, chemically resemble nicotine. They are considered safer for humans than many other classes of pesticides, because they interfere with neural pathways that are more common in insects than in mammals, but from a bee's perspective that obviously isn't much of a recommendation. (The most commonly used neonicotinoid, imidacloprid, is considered "highly toxic" to bees and therefore is not supposed to be applied while they are around.) In March, Hackenberg sent a letter to growers, asking that they "please try to use something beside these products" on their crops.

Meanwhile, he and Davey began to rebuild. They ordered new bees, which they had air-freighted from Australia, and new queens, which were flown in from Hawaii. They split any colony that seemed to be healthy into two and persuaded a firm that usually sterilizes contact lens solution to blast several truckloads of beekeeping equipment with radiation. By the spring, they had managed to restock some two thousand hives. They had also spent more than three hundred thousand dollars.

"Last year, we had just enough to keep going, but not enough to survive on," Davey told me. "It's, like, give us a sign: either wipe the damn things out or tell us what we're supposed to do here. We're just hanging on by the skin of our teeth. If we go through this another year, we'll be flat-broke out of business."

ON A SUNNY SATURDAY this spring, I drove to Betterbee, an apiary supply store in Greenwich, New York, about forty miles north of Albany. When I arrived, the place was crowded with beekeepers and aspiring beekeepers, some from as far away as Maryland, who were queued up in the parking lot. On reaching the front of the line, I was handed a package, much the way you would be handed a loaf of bread, or a pizza. It gave off a slight, insistent buzz.

A package—in beekeeping, this is a precise rather than a generic term—is a container about the size of a shoebox, with wooden sides and wire mesh covering the front and back. It holds about fourteen thousand worker bees and a single queen, who is housed, along with a few devoted attendants, in a tiny cage. The workers and the queen are not from the same original colony, and the queen is kept secluded to give the other bees time to grow accustomed to her odor. The queen cage has a special stopper on one end, made out of a substance called "queen candy." The hope is that by the time the workers are able to eat their way through the candy and liberate the queen they will have accepted her as their leader and will not try to kill her.

Left to their own devices, honeybees construct their hives in cavities, usually in trees and preferably with small openings that face south. They attach their wax combs to the top of the cavity and build downward in parallel sheets lined on both sides with hexagonal cells. These cells are used for all the bees' various needs—to house the young, to store pollen and nectar, and to preserve honey. (To make honey from nectar, bees combine it with special enzymes in their "honey stomachs" and then evaporate out the water.) The key to beekeeping is to persuade a colony to construct its combs where people can get at them. The ancient Egyptians fashioned conical hives out of hardened mud. Since then, hives have been made out of practically every substance imaginable, including clay, stone, logs, bark, wicker, cork, and bamboo. Skeps—bell-shaped straw hives of the sort still popular in cartoons—were widely used in seventeenth-century Europe and were brought to the New World by some of the earliest colonists. (Before the colonists arrived, there were no honeybees anywhere in the Americas.)

Nowadays, nearly all beekeepers use hives of the same basic design, developed a century and a half ago by the Reverend Lorenzo Lorraine Langstroth, a Congregational minister from Philadelphia. Langstroth suffered from severe psychiatric difficulties; attempting to preach his first sermon, he came down with an acute case of what might be called rector's block and was unable to speak. (He referred to this as the start of his "head troubles.") He took up beekeeping in the hope that the outdoor work would clear his mind.

Langstroth's crucial insight—"I could scarcely refrain from shouting 'Eureka!' in the open streets," he wrote of the moment of revelation—was the concept of "bee space." He realized that while honeybees will seal up passageways that are either too large or too small, they will leave open passages that are just the right size to allow a bee to pass through comfortably. Langstroth determined that if frames were placed at this "bee-space" interval of three-eighths of an inch, bees would build honeycomb that could be lifted from the hive, rather than, as was the practice up to that point, sliced or hacked out of it. He patented L. L. Langstroth's Movable Comb Hive in 1852. Today's version consists of a number of rectangular boxes—the number is supposed to grow during the season—open at the top and at the bottom. Each box is equipped with inner lips from which frames can be hung, like folders in a filing drawer, and each frame comes with special tabs to preserve bee space.

I set up my hive at the edge of a small brook that runs through the backyard. Within a day of being installed, my bees—Italians—were hard at work. They could be seen zipping out of the little opening in the front and returning with yellow wads of pollen stuffed into the baskets on their legs. Even my teenage son found the sight of their proverbial busyness hard to resist. On returning home from school, he would lounge against a nearby tree and watch.

• • •

ONE OF THE PEOPLE that David Hackenberg called to tell about his dead hives was Pennsylvania's state apiary inspector, Dennis van Engelsdorp. Van Engelsdorp does not normally keep bees himself, but at the time that I went to visit him, a few weeks ago, he had eight hives in his yard, arranged in a horseshoe. The hives' owner had identified them as suffering from CCD, and van Engelsdorp had brought them home because he was interested in seeing how long it would take for them to collapse entirely. He asked me if I wanted to have a look. We both put on bee gear—a sort of cross between a hazmat suit and a fencing mask. Van Engelsdorp smoked the first hive, then pulled out a frame. To me, it looked perfectly normal, except that, as van Engelsdorp pointed out, there were almost no bees on it.

"If you look here, you can see eggs, so the queen's trying," van Engelsdorp said, passing me the frame. We had neglected to put gloves on, but this didn't seem to matter, because the few bees there were so listless. I could see eggs, which resemble tiny grains of rice, at the bottom of several dozen cells.

"She's cranking," he went on. "Those eggs are on their side, so they're three days old. But this colony's just not building. It's not a factor of there not being enough young bees coming out, but it seems to be a factor of the fact that the adult population is disappearing. So this is an indication where I think this colony will be completely collapsed next week."

And so it went. In the second colony, van Engelsdorp spotted what are known as "supercedure" cells. These cells, which hang off the honeycomb like misshapen peanuts, are a sign that worker bees are trying to produce a new queen. (An ordinary larva grows into a queen if it is fed on a special high-calorie diet of royal jelly.)

"This is another thing we're seeing," van Engelsdorp told me. "The queens don't seem to hold as long. It's sort of like bees coughing, trying to get rid of some illness that they associate with the queen." A third hive was almost completely empty. Van Engelsdorp poked around inside it, looking for signs of wax moths or small hive beetles, insects that prey on weak colonies, but found none. "So whatever is killing the bees, is it also killing the moths or is it just driving them out, or what?" he asked. After going through the rest of the hives, van Engelsdorp decided to conduct an anatomical inspection. He picked up a worker, pinched the end of her abdomen between his fingers, and pulled. A slimy, cream-colored thread emerged. This was the bee's rectum, its kidneys—or Malpighian tubules—and its intestines. He tossed the bee on the ground, and repeated the process with another bee.

Van Engelsdorp, who is thirty-seven, has a bearish build, thinning blond hair, and deep-set blue eyes. He lives in the woods about thirty miles west of Harrisburg, in a one-room cabin with an unheated porch that he sleeps on year-round. Like many people who started to hear from Hackenberg last fall, van Engelsdorp wasn't initially very concerned. He figured that the problem had to do with mites or—much the same thing—with one of the many diseases, like deformed-wing virus, that the mites transmit. (The fact that Hackenberg hadn't found any dead bees was odd, but sick honeybees often leave the hive to expire.) What convinced him otherwise was slicing up some bees that Hackenberg brought from Florida.

Normally, if you cut open a bee its innards, viewed under a microscope, will appear white. Hackenberg's bees were filled with black scar tissue. They seemed to be suffering not so much from any particular ailment as from just about every ailment. "There was just so much wrong with them," van Engelsdorp recalled. "And there weren't any mites."

After more beekeepers began reporting problems, van Engelsdorp started traveling around the country, collecting samples. Some he preserved in alcohol, for his own lab; others he put on dry ice, to be sent out for more sophisticated molecular tests. (Soon, he was so overwhelmed by samples that he had to hire an assistant, whose job consists entirely of conducting bee autopsies.) When the molecular tests were performed, by entomologists at Penn State, they confirmed van Engelsdorp's initial impression. The bees were infected with just about every bee virus known, including deformed-wing virus, sac-brood virus, and black-queen-cell virus, and also by various fungi and bacteria. In addition, genetic analysis revealed the presence of new pathogens, never before sequenced. Such was the level of infection that van Engelsdorp and other researchers concluded that the bees' immune systems had collapsed. It was as if an insect version of AIDS were sweeping through the hives.

ONE EVENING AT around ten o'clock, in the middle of a downpour, my husband heard an odd noise. It sounded to him like clattering, and it seemed to be coming from the general direction of our hive. When he went outside to check, a bear was standing where previously the boxes had been stacked. The frames were scattered in the weeds.

The next morning, thousands of bees were clustered against the base of a nearby tree. Thousands more were lying dead on the ground. I righted the hive and gathered up the frames. Then, using a garden trowel, I spooned as many of the survivors as I could back into the boxes. That night, the bear attacked again. Do bees suffer? I regret to say that I think they do. Those who made it through the second attack spent their days grumpily buzzing around, or huddled pathetically together. Within two weeks, they, too, were dead.

Most beekeeping manuals advise against trying to raise bees on the basis of a manual. Instead, they suggest finding a more experienced beekeeper who lives nearby or joining a local beekeeping organization. The nearest group I could find was the Southern Vermont Beekeepers, and one evening I attended a meeting in the back of a Manchester bookstore. The guest speaker was the Vermont state apiary inspector, and he spent a while talking about colony-collapse disorder, which hadn't been found in Vermont, and then about varroa mites and tracheal mites and American foulbrood and nosema and small hive beetles, all of which had. When he called for questions, the discussion quickly turned to bears. Practically everyone had a story to tell. Ordinary fences, it was agreed, were useless, and even electric ones could be breached. One man said that he draped his electric fence with bacon; this enticed the bear to stick his nose against the wires and get zapped.

Another recommended driving nails through plywood, then laying the plywood around the hive, nail-side up.

"It definitely keeps the bears out," he said of the arrangement.

"It's not too good for the inspector who steps on a nail," the inspector said.

"Get a tetanus shot," a second man suggested.

On my return home, I relayed what I'd learned to my husband. I told him I was opposed to the nail approach, and he said he was opposed to an electric fence. Ultimately, we settled on a third option, not recommended by anyone. We ran a wire cable between two trees, about twenty feet off the ground, and attached a pulley to it. Then we mounted the hive on a platform that could be raised and lowered by rope. Since it was too late in the season to get another package, we ordered what's known as a nucleus hive, or nuc, which, once again, I went to pick up at Betterbee. It came in a plastic-foam container resembling a cooler. Instead of Italians, the bees in the nuc were Carniolans. The queen, who can be hard to spot, because Carniolans are so dark, had been marked with a tiny yellow dot.

For ten days, the aerial hive worked fine. On the eleventh, I went out to check on it and found the boxes on the ground. The frames had all tumbled out, and the bees were buzzing around wildly. I wasn't sure what had happened until I spotted a small black bear a few yards away. Presumably, he had climbed a tree and swatted at the hive until it fell. I chased him away, put on my beekeeping gear, and set the frames back in place. The platform was dangling over my head. While I was trying to figure out my next move, I got stung on the chin. I retreated, and, when I looked up, the bear had returned. He sidled over to one of the boxes I had righted and tried to tip it over. Bees swarmed toward his face. He tumbled back. He tried again. The same thing happened. He sat down. For several minutes, we eyed each other warily. Finally, a tractor came to mow a nearby field and the sound of it drove him off.

I had lost one colony, and now it seemed that I was about to lose another. My husband announced that he had had enough of bees, but, both morally and journalistically, I felt committed. I persuaded him to restring the cable between two trees that were farther apart, and we raised the hive back up. Much to my amazement, this seemed to work.

DR. IAN LIPKIN is the head of the Jerome L. and Dawn Greene Infectious Disease Laboratory at Columbia Health. He is a slight man with sandy-colored hair and a boyishly unlined face. When he was in medical school, Lipkin intended to become an internist, but then he became interested in neurology, which, in turn, led him to molecular biology. "The thing about molecular biology is, it's like a magic trick," Lipkin told me when I went to speak to him one day this summer in his office, in upper Manhattan. "Once you know the trick, you say, 'I should have seen that.'"

Last December, Lipkin received an email from an entomologist at Penn State. The email asked for his help in solving the mystery of CCD. This was two months before CCD began to make news, and Lipkin had no idea what the entomologist was talking about. He wondered whether the email was genuine. "I get a lot of kooks writing me," he explained. Lipkin knew almost nothing about the ailments of invertebrates, and even less about bees, but his lab had done a lot of work on zoonoses—diseases, like avian flu, that originated in animals and have "jumped" species to infect humans. "I decided, Well, why not diseases of insects? It's sort of a natural extension," he told me. He agreed to help and soon was sent some of the bees that van Engelsdorp had collected.

Lipkin subjected the bees to what is called "metagenomic analysis." As he described it to me, the goal of the process is to extract all the genetic information available from a given sample, in this case not just the bees themselves but also the protozoa, bacteria, viruses, and fungi that had been living in them. "It's like if we took you, your shoes, your socks, your sweater, and sequenced everything," Lipkin said. "Then we need to figure out what was shoes, what was socks, and what was you." (Just last year, scientists at the Baylor College of Medicine, in Houston, published the entire honeybee genome sequence, consisting of some two hundred and sixty million base pairs.) At various points,

Lipkin assigned more than a dozen researchers in his lab to work on the project, sometimes seven days a week.

The metagenomic analysis confirmed van Engelsdorp's initial impression. Bees suffering from CCD were infected not with one pathogen but with many, and the most economical explanation was that their immune systems were compromised. This left the central question still to be answered: Why was this happening?

At the time that I spoke to him, Lipkin had just sent off a paper on CCD to a scientific journal. He was reluctant to discuss its contents, for fear of jeopardizing its acceptance, but he did indicate that it contained what he considered to be a breakthrough. One pathogen in particular was, in his words, "highly associated" with CCD.

"My speculation would be that this particular pathogen is a trigger that takes an otherwise borderline population and throws it over the edge," he told me. "I think that's what we're seeing." Lipkin explained that the process of finding the pathogen responsible for an outbreak was "the same whether we're talking about encephalitis or diarrheal disease or hemorrhagic fevers or respiratory disease. You put up a candidate and then try to tear it down. And, if you can't tear it down, it's probably bona fide. That's how we do science." He wouldn't tell me what kind of pathogen he was talking about in the case of CCD, but soon I learned that it was a virus. I also learned that it was suspected that the virus had entered the United States on imported bees.

So far, CCD has been reported in thirty-six states, included California, New York, Texas, Florida, and New Jersey. There are no reliable estimates of how many hives have been wiped out by the disorder, but commercial beekeepers seem to have been particularly hard hit, with some reporting losses of up to 90 percent. A number of beekeeping businesses have already failed; when I met up with David Hackenberg, he showed me a stack of boxes he had bought from an operation that had recently gone under. It's impossible to predict how many businesses will be left after this year. During the summer, when the nectar is flowing, even weak colonies can appear to be healthy. The test will come in the fall, the season when CCD was first discovered.

I asked several entomologists what could be done if, in fact, CCD did turn out to be caused or, to use Lipkin's word, "triggered" by a virus, and got back a wide range of answers. New, resistant strains of honeybees could, at least in theory, be bred. Bees could, once again in theory, be treated with antiviral drugs, or, alternatively, the virus might burn itself out. (A recent paper co-written by van Engelsdorp shows that unexplained honeybee die-offs have occurred in the United States fourteen times in the past hundred years.) Finally, there is the possibility that CCD could keep spreading until people just give up on raising honeybees.

Under this last, worst-case scenario, other pollinators would have to be found to perform the work that honeybees now do. There are thousands of candidates—mostly other species of bees, but perhaps also certain moths or thrips. For some crops, alternative pollinators could well prove more efficient than *A. mellifera;* honeybees don't particularly like squash or pumpkin flowers, for example, while *Peponapis pruinose,* commonly known as the squash bee, does. But the challenges are enormous. Only a fraction of pollinators are generalists, and even fewer are social. Meanwhile, wild pollinators are, by most accounts, in the midst of a crisis of their own.

Last October, just a few weeks before Hackenberg observed the strange symptoms of CCD, the National Research Council issued a report titled "The Status of Pollinators in North America." Fifteen scientists from the United States, Canada, and Mexico had spent a year reviewing the available literature and interviewing experts. They noted that few systematic studies had been done; still, there was plenty of evidence of decline. The Franklin bumblebee, for instance—a mostly black bee native to northwest California and southwest Oregon—was abundant in 1998 but the following year went into a steep and probably terminal decline. None were seen in 2004 and 2005, and only one was found last year. Similarly, the rusty patched bumblebee, once common in New York, hasn't been sighted since 2001. In Britain, where better records have been kept, more than half the native bumblebee species either have become extinct or are facing extinction in the next few decades. Among the many possible contributing factors that the report cited are habitat loss, pesticide use, climate change, and introduced pathogens. May Berenbaum, a professor of entomology at the University of Illinois, chaired the National Research Council panel; she recently characterized CCD as "a crisis on top of a crisis."

"We can't count on wild pollinators, because we've so altered the landscape that many are no longer viable," she said.

As the National Research Council report noted, invertebrate extinctions don't tend to have much "marquee appeal." Yet if it's a bad sign when an ecosystem loses its large mammals, it is probably an even worse sign when it can no longer support its insects. The report put it this way: "Pollinator decline is one form of global change that actually does have credible potential to alter the shape and structure of the terrestrial world."

As for my honeybees, they seem to be doing fine. After their unfortunate fall, I was worried that the queen might have been crushed or perhaps suffocated by her nervous attendants, an accident known as "balling." In that case, the colony would have had to go through the risky exercise of breeding a new queen and getting her mated. But just the other day I lowered the hive, smoked it, and opened the cover. When I pulled out a frame, it was dripping with honey. I could see lots of fat little larvae curled up in their cells, proof that the queen was alive and laying. To my fond, unpracticed eye, it all seemed beautiful. I put the cover on and hoisted the hive back up. As of this writing, it is still there, swaying between the trees.

Originally published in *The New Yorker,*
August 6, 2007.

Update: Still today the cause of colony collapse disorder remains a scientific mystery. Recently, the number of reported cases has declined, but honeybee losses from all causes remain high. In the winter of 2025 they were particularly high—upward of 60 percent of commercial hives failed.

PART TWO

A SENSE OF PLACE

A SONG OF ICE

Greenland Is Melting

NOT LONG AGO, I attended a memorial service on top of the Greenland ice sheet for a man I did not know. The service was an intimate affair, with only four people present. I worried that I might be regarded as an interloper and thought about stepping away. But I was clipped onto a rope, and, in any case, I wanted to be there.

The service was for a NASA scientist named Alberto Behar. Behar, who worked at the Jet Propulsion Laboratory in Pasadena, might be described as a twenty-first-century explorer. He didn't go to uncharted places; he sent probes to them. Some of the machines he built went all the way to Mars; they are orbiting the planet today or trundling across its surface on the *Curiosity* rover. Other Behar designs were deployed on earth, at the poles. In Antarctica, Behar devised a special video camera to capture the first images ever taken inside an ice stream. In Greenland, he once sent a flock of rubber ducks hurtling down a mile-long ice shaft known as a moulin. Each duck bore a label, offering, in Greenlandic, English, and Danish, a reward for its return. At least two made it through.

When Behar died, in January 2015—he crashed his single-engine plane onto the streets of Los Angeles—he was at work on another probe. This one, dubbed a drifter, looked like a toolbox wearing a life preserver. It was intended to measure the flow of meltwater streams. These so-called supraglacial rivers are difficult to approach, since their banks are made of ice. They are often lined with cracks, and usually they end by plunging down an ice shaft. The drifter would float along, like a duck, collecting and transmitting data, so that, by the time it reached a moulin and was sucked in, it would have served its purpose.

Behar was collaborating on the drifter project with a team of geographers at UCLA. After his death, the team carried on with the project, which itself became a kind of memorial. When the geographers picked a supraglacial river to toss the drifters into, they called it the Rio Behar.

I flew up to the Rio Behar in July with several UCLA graduate students and two drifters. My first glimpse of it was out the helicopter window. Its waters were an impossible shade, a color reserved, in other circumstances, only for Popsicles. That fantastic blue was set against a pure and hardly less fantastic whiteness. "Greenland!" the artist Rockwell Kent wrote, after being shipwrecked in an ice fjord. "Oh God, how beautiful the world can be!"

An earlier wave of students had already set up camp. This consisted of one orange cook tent and nine smaller tents, also orange. Beneath the camp, the ice extended more than half a mile. Dotting its surface were perfectly round holes, each an inch or two in diameter and about a foot deep. The holes were filled with meltwater. On this half-solid, half-liquid substrate, staking the tents had proved impossible. The one I was assigned was tied to a quartet of fuel cannisters. "Don't smoke," someone advised me.

A line of yellow caution tape had been strung about fifty yards from the Behar's edge. Anyone venturing beyond that line, I was instructed, had to be tethered. I borrowed a mountaineering harness, clipped in, and made my way to the bank, where the team's leader, Larry Smith, was conferring with a pair of graduate students. By ice-sheet standards, it was a balmy day—around thirty-two degrees—and Smith was wearing canvas work pants; two plaid shirts, one on top of the other; and a red fleece cap that said "Air Greenland."

"Do you hear that?" he asked me. Above the rush of the river, there was a roaring sound, like waves crashing against a distant cliff. "That's the moulin."

Eighteen months after the plane crash, Smith still had trouble talking about Behar. He had brought to the river a half-liter bottle of Coke, which he was carrying in a side pocket of his pants. In the field, he told me, Behar had more or less lived on Diet Coke. He apologized for having to substitute the sugared variety.

Smith twisted open the bottle, drank from it, then handed it around. Each of the students took a few swigs. When Smith got it back, he wrote his email address on the label, with the message "If found, please contact." Then he lofted the bottle into the Behar and we all watched it disappear, floating toward the moulin in the icy blue.

PEOPLE ATTRACTED TO the Greenland ice sheet tend to be the type to sail up fjords or to fly single-engine planes, which is to say they enjoy danger. I am not that type of person, and yet I keep finding myself drawn back to the ice—to its beauty, to its otherworldliness, to its sheer, ungodly significance.

The ice sheet is a holdover from the last ice age, when mile-high glaciers extended not just across Greenland but over vast stretches of the Northern Hemisphere. In most places—Canada, New England, the upper Midwest, Scandinavia—the ice melted away about ten thousand years ago. In Greenland it has—so far, at least—persisted. At the top of the sheet there's airy snow, known as firn, that fell last year and the year before and the year before that. Buried beneath is snow that fell when Washington crossed the Delaware and, beneath that, snow from when Hannibal crossed the Alps. The deepest layers, which were laid down long before recorded history, are under enormous pressure, and the firn is compressed into the ice. At the very bottom there's snow that fell before the beginning of the last ice age, a hundred and fifteen thousand years ago.

The ice sheet is so big—at its center, it's two miles high—that it creates its own weather. Its mass is so great that it deforms the earth, pushing the bedrock several thousand feet into the mantle. Its gravitational tug affects the distribution of the oceans.

In recent years, as global temperatures have risen, the ice sheet has awoken from its post-glacial slumber. Melt streams like the Rio Behar have always formed on the ice; they now appear at higher and higher elevations, earlier and earlier in the spring. This year's melt season began so freakishly early, in April, that when the data started to come in, many scientists couldn't believe it. "I had to go check my instruments," one told me. In 2012, melt was recorded at the very top of the ice sheet. The pace of change has surprised even the modelers. Just in the past four years, more than a trillion tons of ice have been lost. This is four hundred million Olympic swimming pools' worth of water, or enough to fill a single pool the size of New York State to a depth of twenty-three feet.

An ice cube left on a picnic table will melt in an orderly, predictable fashion. With a glacier the size of Greenland's, the process is a good deal more complicated. There are all sorts of feedback loops, and these loops may, in turn, spin off loops and subloops. For instance, when water accumulates on the surface of an ice sheet, the reflectivity changes. More sunlight gets absorbed, which results in more melt, which leads to still more absorption, in a cycle that builds on itself. Marco Tedesco, a research professor at Columbia's Lamont-Doherty Earth Observatory, calls this "melting cannibalism." As moulins form at higher elevations, more water is carried from the surface of the ice to the bedrock beneath. This lubricates the base, which, in turn, speeds the movement of ice toward the ocean. At a certain point, these feedback loops become self-sustaining. It is possible that that point has already been reached.

ACCORDING TO THE *Encyclopedia of Snow, Ice and Glaciers,* glacial ice "behaves as a non-linear visco-plastic material." To put this differently, ice, like water, flows. For reasons that are not entirely understood, ice flows faster in some parts of the ice sheet than in others. Regions where the flow is particularly swift are known as ice streams. The East Greenland Ice-Core Project, EGRIP (pronounced "ee-grip") for short, sits atop one of the longest and widest of these streams, the Northeast Greenland Ice Stream, or NEGIS (pronounced "nay-gis"). This past June, I flew up to EGRIP on a ski-equipped C-130 Hercules, which those in the know call a Herc. The Herc had small rockets—Jet

Assisted Takeoff units, or JATOS—mounted below each wing. The JATOS were there in case it got too hot and the runway at EGRIP, which consists entirely of snow, grew sticky.

EGRIP is run by a Danish glaciologist named Dorthe Dahl-Jensen. Dahl-Jensen is a soft-spoken woman with bright blue eyes and an asymmetrical sweep of white hair. She's fifty-eight and has been working on the ice sheet almost every summer for the past thirty-five years. Initially, as a graduate student at the University of Copenhagen, she'd had to talk her professor, a geophysicist named Willi Dansgaard, into allowing her to come. Dansgaard was against the idea, because the last time he'd brought along a female student the camp's cook had fallen in love with her and stopped cooking. As it happened, on her first trip to the ice sheet, Dahl-Jensen fell in love. She and her husband, J. P. Steffensen, also a glaciologist, have four children. During the summer, they trade off raising the kids and overseeing operations on the ice.

EGRIP is very much a work in progress. Last year's field season was devoted to hauling equipment from a defunct ice station two hundred and seventy-five miles away. This included a whole building, containing a kitchen, a rec room, a bathroom, a dining hall, and an office. The building, which weighs thirty-five tons, was mounted on skis and dragged behind a tractor equipped with extra-heavy-duty treads.

When I arrived, midway into the 2016 field season, construction at EGRIP was still underway. A network of vaulted tunnels had been created, with floors and walls carved out of snow. These glittered from all angles, like something out of *A Thousand and One Nights*. At the bottom of one tunnel a deep pit had been cut using a chain saw, and, next to the pit, a carpenter was erecting a wooden platform. The bricks of ice that had been pulled from the pit had been lugged up to the surface and arranged into what I can only believe is the world's northernmost outdoor bar.

All of this—the tunnels, the pit, the platform—had been fashioned to accommodate an enormous drill, parts of which had traveled with me to EGRIP on the Herc. The point of the project is to send the drill from the top of the ice sheet to the bottom, a distance of more than eight thousand feet. Owing to the way the ice sheet was created, layer upon layer, the drill, as it descends, will, in effect, be boring through history. (In the case of an ice stream, it is possible to step in more or less the same river not just twice but any number of times.) If all goes as planned, Dahl-Jensen told me, the drilling will be complete in 2020. Meanwhile, the ice stream will be moving at the surface, at a rate of around six inches a day, and EGRIP will be moving with it, meaning that the borehole will start to bend. One of the toughest challenges of the project is figuring out how to keep the drill from getting stuck.

THE MAIN BUILDING at EGRIP—the one that got schlepped across the ice—is a sort of double geodesic dome, with one dome resting on the other, like the lid on a casserole. At the very top of it there's a cupola. The domes and the cupola are covered in rubber sheeting, and to my eye the whole arrangement resembled a big black time bomb.

My second day at EGRIP, everyone gathered in the double dome for what was billed as the "first-ever" master's thesis defense on the ice. The chairs in what normally served as the rec room had been rearranged, classroom style, and one of Dahl-Jensen's students, a bearded young man named Kristian Høier, rose to discuss the issue of "surface buckling." Although Høier spoke in English, I couldn't understand most of his presentation, which turned on details of the equations he'd used in his mathematical model. He seemed nervous and kept sighing loudly, which I also couldn't understand, as it was obvious that the first-ever thesis defense on the ice was going to result in the first-ever pass. When his presentation was over, Dahl-Jensen opened a case of champagne and everyone put on parkas and heavy boots to stand around the outdoor bar. It was evening, but, since the sun never sets in northeastern Greenland in June, still bright. The snow, flat and unbroken in all directions, had acquired a bluish tint. Dahl-Jensen offered a toast to Høier, who seemed intent on getting hammered as quickly as possible. I left my cup on the bar and went back into the building to get my camera. By the time I returned, my drink was halfway to a champagne slushie.

As its name suggests, the NEGIS flows in a northeasterly direction. It has its head, as it were, at the center of Greenland, near the highest point on the ice sheet. Its mouth empties into the Fram Strait. There icebergs the size of city blocks split off, or, as geologists say, calve, and float away. Given enough time, EGRIP, like some drifting barge, will also reach the Fram and topple in.

All over Greenland, ice streams like the NEGIS are picking up their pace. In the process, they are dumping more and more ice directly into the oceans. Currently, it's estimated that Greenland is losing about as much ice from calving as it is from melt. One group of scientists argues that, of the two forms of loss, melt is the more worrisome, as, in a warming world, it must increase. But the behavior of ice streams is less well understood, and some scientists argue that, for this very reason, increased calving is potentially even more of a threat.

"The fastest way to get rid of an ice sheet is to throw it into the ocean" is how Sune Olander Rasmussen, the field office manager for EGRIP, put it to me.

"The ice streams have really, really surprised us," Dahl-Jensen said. "To drill down into an ice stream and see: How does it actually flow? How much is it sliding? How much is it melting at the bottom? I see that as the most important goal of this project."

Once an ice stream starts to accelerate, it may be impossible to stop. "In some cases, you have, in theory, this irreversible process," Kerim Nisancioglu, a climate scientist from the University of Bergen who works at EGRIP, told me. "And you set it off and it just goes. It drains.

"This system is huge," Nisancioglu continued, referring to the ice stream we were standing on. "It has a lot of water to drain. So it could keep going for a long time. How far can it go? Will it keep accelerating indefinitely until it runs out of ice? This is unknown." All on its own, the NEGIS has the potential to raise global sea levels by three feet.

THE FIRST ATTEMPT to drill through the Greenland ice sheet was made in the early nineteen-sixties at a U.S. Army outpost called Camp Century. Some fifty years later, the camp remains far and away the biggest thing ever built on—or, really, under—the Greenland ice. Camp Century had a bar, a chapel, a barbershop, a movie theater, and a nuclear reactor. All were housed in a network of snow tunnels like those at EGRIP, but extending for miles. The ostensible purpose of the base was to promote Arctic science, but in the nineteen-nineties an investigation by the Danish government revealed this to be a ruse. What the army had really been up to was developing a new system for storing intercontinental ballistic missiles. Its plan was to install a sub-glacial railway and shuttle ICBMs around in a Cold War shell game. The code name for the scheme was Project Iceworm.

The drilling at Camp Century was not exactly a secret; still, visitors were not allowed to watch while it was underway. It yielded hundreds of cylinders of ice, each about a yard and a half long and four inches in diameter. These sat around in a freezer in New Hampshire until Willi Dansgaard, Dahl-Jensen's teacher, got hold of them.

Dansgaard, who died in 2011, was an expert on the chemistry of precipitation. Presented with a sample of rainwater, he could, on the basis of its isotopic composition, determine the temperature at which the precipitation had formed. This method, he realized, could also be applied to snow. Dansgaard was able to read the Camp Century core as a sort of almanac of Greenlandic weather. He could tell how the temperature had changed ice layer by ice layer, which is to say year by year.

Mostly, Dansgaard's results confirmed what was already known about climate history. For instance, he observed that Greenland had experienced a cold snap from around the year 1300 to 1800—the so-called Little Ice Age. He found that for most of the past ten thousand years it had been relatively warm on Greenland, and for tens of thousands of years before that it had been frigid.

But Dansgaard also turned up something totally unexpected. It appeared from his analysis of the Camp Century core that, in the midst of the last ice age, temperatures on Greenland had shot up by fifteen degrees in fifty years. Then they'd dropped again, almost as abruptly. This had happened not just once but many times.

Everyone, including Dansgaard, was perplexed. A temperature swing of fifteen degrees? It was as if New York City had suddenly become Houston or Houston had become Riyadh. Could these violent swings in the data correspond to real events? Or were they some sort of glitch?

Over the next forty years, five more complete cores were extracted from different parts of the ice sheet. Each time, the wild swings showed up. Meanwhile, other climate records, including pollen deposits from a lake in Italy, ocean sediments from the Arabian Sea, and stalactites from a cave in China, revealed the same pattern. The temperature swings became known, after Dansgaard and a Swiss colleague, Hans Oeschger, as Dansgaard-Oeschger events. There have been twenty-five such events in the past hundred and fifteen thousand years.

Ice ages are triggered by small, periodic changes in the earth's orbit that alter the amount of sunlight hitting different parts of the globe at different times of year. The Dansgaard-Oeschger (or D-O) events, which occurred at irregular intervals, have no apparent cause. The best explanation anyone has been able to offer is that the sheer complexity of the climate system renders it unstable—capable of flipping from one state to another.

"It's a great interplay between the glaciers, the atmosphere, the sea ice, and the oceans," Dahl-Jensen told me. We were sitting in her office, which is in the cupola of the double dome and reachable, tree-house style, via ladder. It was a few hours after the thesis defense, and the sun was finally dipping toward the horizon.

"But we still struggle to understand how we can get these very big abrupt changes," she went on. "And I really think that understanding them is one of the most important challenges we face. Because if we fail to be able to understand them in our past, we don't have the tools to understand the risk of them in the future."

All the D-O events predate the emergence of civilization, and this is probably no coincidence. In climactic terms, the past ten thousand years have been exceptionally stable. Go back further than that, and devastating shifts show up again and again. Somehow or other, our ancestors came through that chaos, but before the invention of agriculture people traveled light. They never stayed in one place long enough to develop complex societies and all that followed—cities, metallurgy, livestock, writing, money. When a D-O event occurred, bands of hunter-gatherers presumably picked up and moved on. Either that or they died out.

|| • • • • •

GREENLAND IS THE world's largest island, unless you count Australia, which is usually put in its own category, since it's a continent. The ice sheet covers about 80 percent of the island, making it one of the least green places on earth.

"Greenland should be called Iceland and Iceland should be called Greenland," Inuuteq Holm Olsen, Greenland's representative to the United States, told me, with a shrug of irritation. "You don't know how many times I've heard that." If Greenland were its own country, it would be the biggest nation in Europe, although, geologically speaking, it's part of North America. The ice-free territory alone—some hundred and seventy thousand square miles—is larger than Germany. As it is, the island is ruled by the Kingdom of Denmark, and Olsen occupies an office in the basement of the Danish Embassy, in Washington, D.C. Like most Greenlanders, he's of Inuit descent.

For as long as they could, the Danes kept Greenland under a sort of reverse quarantine: the goal was not to keep residents in but everybody else out. Foreigners wishing to visit had to apply to Copenhagen for approval; the difficulties of obtaining permission, Rockwell Kent complained in 1930, were "serious and many." (At that point, there was no such thing as private property on the island, and, indeed, even today, in keeping with Inuit tradition, all land is held in common.) According to the Danes, the arrangement was maintained for the good of the Greenlanders, to guard them against the "destructive trends" of modern life. As late as 1940, many families still lived in turf houses and lit their homes with seal-oil lamps.

During the Second World War, Denmark was occupied by the Nazis, and the United States built several air bases on Greenland. By the time the conflict was over, Greenlanders had seen too much of modern life, destructive or otherwise, to go back. What followed was what one Danish chronicler has described as "a social quantum leap unmatched in depth, extent and pace anywhere in the world."

Today, Greenland has fifty-six thousand residents, twelve thousand internet connections, fifty farms, and, by American standards, no trees.

(The native dwarf willows top out at about a foot.) One Greenlander I met, who'd recently left the island for the first time to attend a meeting in upstate New York, told me that his favorite part of the trip had been the noise of the wind sighing through the leaves.

"I love that sound," he said. "*Shoosh, shoosh.*"

There are few roads in Greenland—to get from one town to another you have to take a boat or fly—and, aside from fish-processing plants, little industry. A block grant of five hundred and thirty-five million dollars, sent every year by the Danes, constitutes nearly a third of the island's GDP. In a measured, Scandinavian sort of way, relations between the grantor and the grantee are tense. In 2008, Greenlanders voted overwhelmingly in favor of moving toward independence. Under what's known as the self-rule agreement, which was approved in Copenhagen and in the Greenlandic capital of Nuuk, Greenland gained the right to negotiate some of its own foreign agreements—hence Olsen's basement office. Greenlandic, an Inuit dialect, became the island's official language, and the size of the annual grant from Copenhagen was capped.

Greenland celebrates its version of July 4 on June 21. This past June, in an effort to demonstrate solidarity, the Danish government instructed its agencies and embassies to raise the Greenlandic flag. A half-red, half-white circle on a half-white, half-red background, the flag is supposed to represent the ice sheet over the ocean, with the sun sinking beneath the waves. Many Danish agencies complied with the directive but, awkwardly enough, flew the flag upside down.

"We have a lot of postcolonial problems," Niviaq Korneliussen, a twenty-six-year-old woman who may be Greenland's most widely read novelist, told me. "We have a lot of racism going on from both ends. There are a lot of young people who hate Danish people because their parents did. So there's a long way to go for things to get better."

ALMOST A THIRD of the island's population lives in Nuuk, which is by far Greenland's largest town, and, in between trips onto the ice sheet, I went for a visit. On my ten-minute taxi ride from the airport, I think I passed through all three of Greenland's stoplights.

Nuuk sits on the southwest coast. It was founded in the early eighteenth century by a Danish-Norwegian missionary named Hans Egede and for most of its existence was known as Godthåb. When Egede arrived, he discovered that the native people had neither bread nor a word for it, so he translated the line from the Lord's Prayer as "Give us this day our daily seal." Today, a giant statue of Egede presides over Nuuk much the way Christ the Redeemer presides over Rio.

My visit to Nuuk coincided with a political conclave hosted by Greenland's largest labor union. Many of the island's elected officials were supposed to be there, so one afternoon I made my way over. The walk took me past a set of ten identical Soviet-style apartment complexes. These were put up in the nineteen-sixties, when the Danes decided to empty many of Greenland's tiny fishing villages and concentrate people in larger towns. In their day, the apartments, with electricity and indoor plumbing, seemed the height of modernity; now, surrounded by sleeker, newer buildings, they're considered a slum.

The conclave was being held at a large gym with a vaulted ceiling. Inside, about a hundred people were listening to a panel discussion on the subject "Is Greenland ready for the mineral industry?" Simultaneous translation was being provided from Greenlandic into Danish, from Danish into Greenlandic, and from both languages into English. I picked up a headset, but the English channel kept cutting out, and after a while it occurred to me that I was probably the only person trying to listen to it. Tables had been set up around the perimeter of the gym; from them the island's political parties were dispensing sweets, pamphlets, and swag. Groups of impossibly cute children were roaming from one table to the next, grabbing as many balloons and cookies as they could. I struck up a conversation with a man named Per Rosing-Petersen, who was staffing the table for a party called Partii Naleraq. (Almost all Greenlanders nowadays have Danish names, and, owing to hundreds of years of intermarriage, many also have blue eyes.) It turned out that Rosing-Petersen was a member of the Greenlandic parliament. Partii Naleraq's offerings included orange plastic bracelets that said *Tassa asu! Naalagaajinngorta!,* which he translated as "Let's go! Independence!"

"If you look at the businesses in Greenland, 90 percent are owned and managed by Danes," Rosing-Petersen told me. "The Greenlanders are the working class. I call it apartheid—de facto apartheid. We want to change this picture."

Though Greenland's independence movement has nothing directly to do with climate change, indirectly the links are many. For Greenland to break away, it would have to sacrifice the annual grant from Denmark, which would leave a gaping hole in its budget. The island is rich in minerals, and the theory is that these will become easier to get at as winters grow shorter and harbors remain ice-free year-round. Greenland's deposits of rare earth elements are, by some accounts, the largest outside China; the island also has significant deposits of iron, zinc, molybdenum, and gold. In 2014, the Greenlandic government released a plan that called for at least three new mines to be operating within four years. "The mineral resources should—so to speak—be made to work for us," the plan said.

Next to Partii Naleraq's was the table for Siumut, Greenland's ruling party. Manning it was another member of Parliament, Jens-Erik Kirkegaard, who, as it happened, had been the minister of industry and mineral resources when the plan was released.

"We haven't had that boom yet," Kirkegaard acknowledged. At the time I visited, the island had no working mines, and the only one under construction—a ruby mine south of Nuuk—was stalled because its Canadian backers had run out of cash. Mostly Kirkegaard blamed the collapse in commodity prices.

"A few years back, mineral prices were very high, but then they declined very hard," he told me. Still, he was optimistic. More melt off the ice sheet meant more attention for Greenland. "Climate change does a lot of marketing for us," he said. "It's easier to attract investment." And as the shipping season grew longer, costs would come down: "Some projects that weren't economical, maybe they will be as conditions change."

• • •

GREENLAND'S INSTITUTE of Natural Resources, known in Greenlandic as the Pinngortitaleriffik, occupies a stylish wood-and-glass complex at the edge of Nuuk. The day after the conclave at the gym, I went to the institute to speak to Lene Kielsen Holm, a social anthropologist who studies Greenlanders' perceptions of climate change. Holm does a lot of her work in Qaanaaq, a town in Greenland's northwest corner that was founded in the early nineteen-fifties, when the United States decided to expand one of its air bases—Thule—and forced most of those living in the area to move out of the way. Qaanaaq, population six hundred and thirty, is one of the few places in Greenland where people still subsist on what they catch. "They have always been adapting to a changing environment," Holm said of the hunters and fishermen she interviews. "This is their daily life. If they didn't have this kind of know-how, they wouldn't survive.

"I think it's part of our culture that we have been living with changes for a long time," she added.

That Greenlanders are unusually resilient is a view I heard many times. "Denmark will disappear," Rosing-Petersen told me. "Holland will disappear. But Greenland will still remain. We've been adapting to living conditions for five thousand years."

Certainly, it's true that life in Greenland is tough. In Qaanaaq, during the winter months, temperatures average around ten degrees below zero and the sun never appears above the horizon. "When the long Darkness spreads itself over the country, many hidden things are revealed, and men's thoughts travel along devious paths," a west Greenlander told the explorer Knud Rasmussen sometime around 1904.

But the record of human habitation of Greenland testifies to more than human resourcefulness. Depending on how you count, Greenland has been a graveyard for four, five, or even six societies.

The first people to migrate to Greenland are known as the Independence I. This group made its way to the island, probably from Canada, about forty-five hundred years ago and settled in a particularly inhospitable territory some four hundred miles northeast of where EGRIP sits today. The *Atlas of the North American Indian* notes that the Independence I people "lacked two elements later Arctic dwellers would consider essential: adequate clothing and reliable fuel for fire in a treeless landscape." Somehow, they managed to eke out a living for almost a millennium. Then they disappeared.

The Independence I people were followed by a group called Independence II, which also vanished. Meanwhile, people known as the Saqqaq arrived in western Greenland. They lasted almost two thousand years and were replaced by what archaeologists call the Dorset. Recent DNA analysis of their remains suggests that both the Saqqaq and the Dorset died off without descendants. From around the time of the birth of Christ to around the time of Charlemagne, Greenland was, it appears, uninhabited.

In the late tenth century, the island was repopulated, this time from the east, by a contingent of Norse led by Erik the Red. It's debated whether Erik called the place Greenland because at that time it really was greener or because he thought it would be good PR. The Norse established two main colonies: the Western Settlement, which was not far from present-day Nuuk, and the Eastern Settlement, which was actually in the south. The settlements prospered and grew until something went terribly wrong. When Hans Egede set out for Greenland, in 1721, he was hoping to bring Protestantism to the Norse, who, he worried, had missed out on the Reformation. But all that was left of the settlements was ruins.

Archaeologists have since determined that the Western Settlement failed around the year 1400 and the Eastern Settlement a few decades later. In climatological terms, this timing is suggestive. The Europeans arrived in Greenland during the so-called Medieval Warm Period, and they vanished not long after the onset of the Little Ice Age.

Still, archaeologists have sought alternative explanations for their disappearance. It's been hypothesized that the Norse were overpowered by the Inuit, who arrived in Greenland, also from Canada, sometime around A.D. 1200, or that they were done in by a drop in the value of walrus ivory. In *Collapse,* Jared Diamond attributes their demise to an oddly self-punishing cultural conservatism. The European settlers had brought with them cattle and sheep. According to Diamond, they continued to rely on their livestock even though they would have been a lot better off copying the Inuit and adopting a marine-based diet. "The Norse starved in the presence of abundant unutilized food resources," he writes. But, according to more recent research, based on the isotopic composition of Norse bones, the Europeans did ditch their cows. By the time the Norse vanished, at least half their calories were coming from seal meat.

"If anything, they might have become bored with eating seals" is how Niels Lynnerup, of the University of Copenhagen, one of the scientists who led the research, put it.

"It's one of those things where, wow, you realize you can be resilient, you can be adaptive, you can be clever, and you can still all be extinct," Thomas McGovern, a professor of archaeology at Hunter College, who has studied the Norse for thirty-five years, told me.

As Greenland warms, the record of the Norse settlements, along with any clues that it might yield, is being erased. "Back in the old days, these sites were frozen most of the year," McGovern continued. "When I was visiting south Greenland in the nineteen-eighties, I was able to jump down in trenches guys had left open from the fifties and sixties, and sticking out the sides you could see hair, feathers, wool, and incredibly well-preserved animal bones." A graduate student of McGovern's who started working in Greenland in 2005 found at the same sites mostly decomposing mush.

"We're losing everything," McGovern said. "Basically, we have the equivalent of the Library of Alexandria in the ground, and it's on fire."

III ● ● ● ● ●

THE TOWN OF Ilulissat sits three hundred and fifty miles north of Nuuk, above the Arctic Circle. It's home to one of Greenland's richest archaeological sites—a stretch of springy tundra that was inhabited first by the Saqqaq, then by the Dorset, and finally by the Inuit. Near the abandoned settlement is a bare stone ledge overhanging a fjord. Elderly Greenlanders used to jump from the ledge to avoid becoming a burden to their families, or so the story goes. The day I went to stand on the ledge, several Danish tourists were taking photos and batting away mosquitoes. Instead of jumping, we had come to admire the view.

Rising from the fjord in front of us was a vast, improbable collection of icebergs. These were jammed together, as if in a frozen metropolis. Towers of ice leaned against arches of ice, which pressed into palaces of ice. Some of the icebergs had smaller icebergs perched on top of them, like minarets. There were ice pyramids and what looked to me like an ice cathedral. The city of ice stretched on for miles. It was all a dazzling white except for pools of meltwater—that fantastic shade of Popsicle blue. Nothing moved, and, apart from the droning of the mosquitoes, the only sound was the patter of water running off the bergs.

The suicide ledge is a good place to go to feel small—presumably that's why it was chosen. Standing at its edge, I could imagine how the Saqqaq and the Dorset were awed by the inhuman beauty. But today even sublimity has been superseded.

The city of ice is the product of the Jakobshavn ice stream. Like the NEGIS, the Jakobshavn originates in central Greenland, only it flows in the opposite direction and into a long fjord. Where the ice meets the water, there's a calving front, and it's here that the ice arches and ice castles take form. These float down the fjord toward Ilulissat. (The town's name is Greenlandic for "icebergs.") They would continue on out to sea, except that they're blocked by a submarine ridge—a moraine—composed of rocky debris left behind when the ice sheet shrank at the end of the last ice age. The biggest icebergs become lodged on the moraine and the smaller ones crowd in behind, as in a monumental traffic jam. The very largest, which weigh upward of a hundred million tons, can hang around for years before slimming down enough to float free. (It is believed that one of these liberated giants from Ilulissat was the iceberg that sank the *Titanic*.)

Eight thousand years ago, the Jakobshavn filled the fjord completely, all the way to the moraine. By the mid-nineteenth century, when the first observations were recorded, the position of the calving front had shifted inland by about ten miles. Over the next hundred and fifty years, the front's position shifted again, by another twelve miles.

Then, suddenly, in the late nineteen-nineties, the Jakobshavn's stately retreat turned into a rout. Between 2001 and 2006, the calving front withdrew nine miles. Just in the past fifteen years, it has given up more ground than it did in the previous century. The fjord extends for at least another forty miles and deepens as it moves inland. At this point, there doesn't seem to be anything to prevent the calving front from withdrawing the entire way.

"It appears now that the retreat cannot be stopped," David Holland, a professor at NYU who studies the Jakobshavn using seals equipped with electronic sensors, told me. (When the seals surface after a dive, the sensors transmit data about conditions in the fjord.)

Meanwhile, as the calving front has receded, the ice stream has sped up. This appears to be the result of yet another feedback loop. Since the nineties, the Jakobshavn has nearly tripled its pace. In the summer of 2012, it set what's believed to be an ice-stream record by flowing at the distinctly unglacial rate of a hundred and fifty feet per day, or more than six feet an hour. The Jakobshavn's catchment area is smaller than the NEGIS's; still, there's enough ice in it to raise global sea levels by two feet.

A LOT OF ILULISSAT is given over to dogs. They have their own neighborhoods—large expanses of dust and rock, where they live chained up around industrial-size vats of water. In my walks around town, I encountered three dog settlements that spread over several acres, and behind my hotel there was a small satellite encampment. In the endless summer sun, the dogs looked stricken. They lay around, panting under their thick coats. Occasionally, one group would start to bay and then the rest would take up the cry, so that the whole town seemed to be howling.

Ilulissat's dogs are all the same kind, a particularly cold-hardy breed of husky, which the Inuit brought with them when they migrated to Greenland. To maintain the purity of the breed, no other type of dog is allowed north of the Arctic Circle.

The huskies used to be central to Greenlandic life. "Give me dogs, give me snow, and you can keep the rest," Knud Rasmussen, the explorer, who was born in Ilulissat in 1879, supposedly once said. As recently as 1995, Ilulissat, a town of some forty-six hundred people, was home to more than eight thousand dogs. In the past twenty years, the canine population has crashed. Now there are only about two thousand dogs. This, too, is an index of global warming.

Ole Dorph, Ilulissat's mayor, works out of a corner office in the town's surprisingly sprawling city hall. He's sixty-one, with a craggy face and rectangular glasses. Dorph grew up in Ilulissat, and he told me that, when he was a child, every year the town was iced in from November to April. During those months, residents used their dog sleds to go fishing and seal hunting.

"In the old days, you could take your sled and go to Disko Island," Dorph said. The island, the largest in Greenland outside of Greenland itself, lies about thirty miles west of Ilulissat, across Disko Bay.

Since no supply ships could get into Ilulissat's harbor, for six months a year residents had to live off whatever provisions the stores had laid in, plus whatever they caught. When the ice broke up in the spring and the first ship arrived, "everyone was very happy," Dorph recalled. "We could buy new apples." To announce the boat's approach, the town would "shoot off a cannon three times—*bang, bang, bang.*"

Then, in the nineties, the bay started to freeze later and later until, finally, it didn't freeze at all. "The last time we had ice we could use was in 1997," Dorph told me.

The loss of ice cover from Disko Bay is part of the general decline in Arctic sea ice—a decline that's been so precipitous it now seems likely there will be open water at the North Pole in summer within the next few decades. Since sea ice reflects the sun's radiation and open water absorbs it, the loss has enormous implications for the planet as a whole. (Sea ice doesn't contribute to sea-level rise, because it floats, displacing an equivalent amount of water.) Locally, in Ilulissat, the most obvious impact has been on transportation. Once the bay stopped freezing, supply ships could arrive in January, and sleds became obsolete. Dogs no longer seemed worth the seal meat it took to feed them. Many were euthanized. Those that remain are used mostly for sport.

Dorph told me that people in Ilulissat were "sad because our dogs are going down" but that this unhappiness was more or less balanced by the benefits of open water. Ilulissat's major source of income is halibut, and its small harbor, which sits on the opposite side of town from the fjord, is crowded with fishing boats.

"The fishermen, they can take their boats out in winter," Dorph said. "They feel it's okay. The price of fish is going up, so the fishermen, they have good days." I was reminded of what I'd heard in Nuuk—that climate change, while regrettable in many ways, was, for Greenlanders, filled with economic promise. I asked Dorph, a member of the ruling Siumut Party, about independence.

"I hope it will happen in maybe ten or twenty years," he said. "It's our key to growing up."

ONE EVENING while I was staying in Ilulissat, I hired a boat to go up the coast. The owner, who was also the captain, was a Dane named Anders Lykke Laursen. He met me at the harbor wearing a pair of yellow-tinted sunglasses, which, he explained, would help him spot dangerous chunks of floating ice. The boat, he went on to assure me, had a double hull and met all the standards the Danish Maritime Authority had laid out for operating in the Arctic. If we did hit some ice, he advised, "it's going to sound bad—but don't worry."

About ten miles north of Ilulissat, we passed the tiny town of Oqaatsut, a collection of bright-painted houses hugging the rocks. (*Oqaatsut* is Greenlandic for "cormorants.") From the boat not a soul was visible, but when I looked it up later in the phone book—there's one edition of the white pages for all of Greenland, and it's about an eighth of an inch thick—I found that Oqaatsut had eighteen listed numbers. We motored on, dodging refrigerator-size blocks of floating ice as well as several massive icebergs that had broken free from the moraine. Beyond Oqaatsut, the coast rose up. A thin waterfall hundreds of feet high twisted off the rocks. Almost anywhere else in the world, the falls would have been a major tourist attraction; in the great emptiness of west-central Greenland, it didn't even have a name.

Finally, after about three hours, we came within sight of our destination, a rock-strewn cove. It also had no name; its coordinates—69.868245N by 50.317827W—had been sent to me by Eric Rignot, a glaciologist from the University of California, Irvine. The cove was shallow, so we paddled ashore in a rubber dinghy, pushing ice chunks out of the way with the oars.

Rignot, who grew up in France, studies both Greenland's ice sheet and Antarctica's. Two years ago, he published a paper arguing that a key section of the West Antarctic ice sheet, the Amundsen Sea sector, had gone into "irreversible retreat." The Amundsen Sea sector contains more than two hundred thousand cubic miles of ice, meaning that, if Rignot's analysis is correct, it will, inevitably, raise global sea levels by four feet.

"THIS IS WHAT A HOLY SHIT MOMENT FOR GLOBAL WARMING LOOKS LIKE," *Mother Jones* declared when the paper was released.

Rignot and three of his students had set up camp on a steep hill just beyond the beach—a cluster of pup tents facing a glacier-filled fjord. In the slanted sunlight—it was about nine p.m.—the glacier, known as Kangilernata, seemed to be glowing. Its calving front, a hundred-and-thirty-foot vertical wall of ice, appeared upside down in the milky-blue waters of the fjord. Behind it, ice stretched to the horizon. Again, I was hit, and vaguely sickened, by Greenland's inhuman scale.

Rignot and his students were monitoring Kangilernata's movements with a portable radar set, which resembled a rotating badminton net. "We measure changes in the condition of the glacier within millimeters," Rignot told me. "It's like making a movie of the flow." But even without sophisticated equipment the glacier's retreat was apparent. Rignot pointed to a fifty-foot-wide band of gray along the walls of the fjord. This showed how much Kangilernata's height had fallen. Coal-black moraines marked the retreat of its calving front. In the past fifteen years, the front has pulled back two miles.

Kangilernata is what's known as a marine-terminating glacier. So is Jakobshavn, and so, too, are most of the glaciers in West Antarctica. This means that they have one foot in the water and, as the world warms, are melting from the bottom as well as from the top. NASA is so concerned about this effect that it has launched a research project called, suggestively, Oceans Melting Greenland, or OMG. (Rignot is one of the principal investigators on the project.)

At Kangilernata, the team was measuring the water temperatures at the base of the calving front every other day. This involved taking a Zodiac into the fjord, dangling some instruments over the side, and hoping the boat wouldn't be swamped by falling ice.

"What concerns me the most is that this is the kind of experiment we can only do once," Rignot said. "A lot of people don't realize that. If we start opening the floodgates on some of these glaciers, even if we stop our emissions, even if we go back to a better climate, the damage is going to be done. There's no red button to stop this."

I FIRST VISITED the Greenland ice sheet in the summer of 2001. At that time, vivid illustrations of climate change were hard to come by. Now they're everywhere—in the flooded streets of Florida and South Carolina, in the beetle-infested forests of Colorado and Montana, in the too-warm waters of the mid-Atlantic and the Great Lakes and the Gulf of Mexico, in the mounds of dead mussels that washed up this summer on the coast of Long Island and the piles of dead fish that coated the banks of the Yellowstone River.

But the problem with global warming—and the reason it continues to resist illustration, even as the streets flood and the forests die and the mussels rot on the shores—is that experience is an inadequate guide to what's going on. The climate operates on a time delay. When carbon dioxide is added to the atmosphere, it takes decades—in a technical sense, millennia—for the earth to equilibrate. This summer's fish kill was a product of warming that had become inevitable twenty or thirty years ago, and the warming that's being locked in today won't be fully felt until today's toddlers reach middle age. In effect, we are living in the climate of the past, but already we've determined the climate's future.

Global warming's back-loaded temporality makes all the warnings— from scientists, government agencies, and, especially, journalists— seem hysterical, Cassandra-like—*Otototototoi!*—even when they are understated. Once feedbacks take over, the climate can change quickly, and it can change radically. At the end of the last ice age, during an event known as meltwater pulse 1A, sea levels rose at the rate of more than a foot a decade. It's likely that the "floodgates" are already open and that large sections of Greenland and Antarctica are fated to melt. It's just the ice in front of us that's still frozen.

On my last day in Ilulissat, I decided that, since I might not be coming back, I ought to go see the ice city again. My hike took me through one of the dusty dog encampments and by the town's old heliport, where, to help boost tourism, a Danish philanthropy is planning to erect a viewing platform overlooking the fjord. (The platform "will provide a front-row seat for the melting ice sheet," the head of the philanthropy said in June, when the winning design was announced.) The ice city didn't appear to have changed much, and I recognized some of the same arches and castles I'd seen earlier. It was a cloudless morning, and again, apart from the mosquitoes, nothing was moving. I'd brought along a notebook and started to make a list of the shapes before me. One iceberg reminded me of an airplane hangar, another of the Guggenheim Museum. There was a sphinx, a pagoda, and a battleship; a barn, a silo, and the Sydney Opera House.

To get back to town, I followed a different route. This one took me past Ilulissat's cemetery. The plots were marked with white wooden crosses and heaped with bright-colored plastic flowers. It was a lovely and oddly cheerful sight, the graveyard overlooking the ice.

Originally published in *The New Yorker*,
October 24, 2016.

Update: Since my visit to Greenland in 2016, the ice sheet has lost an additional two and a half trillion tons of ice. Even so, mining there remains extremely challenging. At the start of 2025, there was only one active mine in the country, which—for now, at least—remains part of the Kingdom of Denmark.

THE LOST CANYON

Drought Is Shrinking Lake Powell,
Revealing a Hidden Eden

L AKE POWELL, which some people consider the most beautiful place on earth and others view as an abomination, lies in slickrock country, about two hundred and fifty miles south of Salt Lake City. Not long ago, I made the trip from Salt Lake to Powell in a rental car. The drive wound by Orem and Provo, then through a landscape so parched that even the sagebrush looked thirsty. A few miles shy of the lake, in the nearly nonexistent town of Ticaboo, I passed a lot where dry-docked cabin cruisers rose, mirage-like, from the desert.

It was the tail end of a record-breaking heat wave and two decades into what's sometimes called the Millennium Drought. When I got to Bullfrog, on the lake's western shore, it was almost six p.m. The car's thermometer read a hundred and twelve degrees. At the Bullfrog marina, families were lugging coolers onto houseboats. Some of the boats had water slides running off the back; others were trailing Jet Skis. Despite the intense heat, the atmosphere was festive. I met a woman who told me that she was using an inheritance to take two dozen relatives out on the lake on the biggest houseboat she could rent—a seventy-five-footer.

"I really shouldn't tell you how awesome it is, because I don't want people from New York to know," she said.

Lake Powell isn't actually a lake; it's an invention of the U.S. Bureau of Reclamation. In the early nineteen-sixties, the bureau erected a seven-hundred-and-ten-foot-tall concrete arch dam on the Colorado

River, near where it crosses from Utah into Arizona. The bureau named the dam for the stretch of the river that it was submerging—Glen Canyon. Behind the dam, water backed up for almost a hundred and ninety miles, creating a reservoir with the shape of a snake that's swallowed a porcupine. At full capacity, Lake Powell stores twenty-four million acre-feet of water, enough to flood the entire state of Massachusetts hip-deep.

In the six decades since the dam was built, the living memory of Glen Canyon has mostly been lost. Relatively few people visited the canyon when it could still be run by raft, and all but a handful of them are now dead. In the meantime, the place has acquired an almost mythical status. It was a kind of Eden, more spectacular than the Grand Canyon and, at the same time, more peaceful. It was a fairy-tale maze of side canyons, and side canyons with their own side canyons, each one offering a different marvel. Edward Abbey, who was one of several writers and artists to float through Glen Canyon shortly before its inundation, called the closing of the dam's gates a "crime." To grasp the nature of this crime, he wrote, "imagine the Taj Mahal or Chartres Cathedral buried in mud until only the spires remain visible."

I first encountered Glen Canyon in a book. It may have been a volume of Eliot Porter photographs, *The Place No One Knew*, or perhaps Abbey's *Desert Solitaire*—I can't remember anymore which I read first. I fell for the myth pretty hard. The wind-sculpted cliffs and sandstone arches of Porter's images, the grottoes, hanging gardens, and amphitheaters big enough for "God's own symphony orchestra" described by Abbey—it seemed heartbreaking that all this was lost. The reservoir—Lake Foul, to its detractors—would, I assumed, last far longer than I would. There was no way I was going to get to see what lay beneath it.

It turns out I was wrong. This isn't because I was too pessimistic; rather, I wasn't pessimistic enough.

In June, Utah's governor, Spencer Cox, announced that the state was so short of water that the only thing that could help would be "divine intervention." He asked citizens of all faiths to join him in praying for precipitation. "We need more rain and we need it now," he said.

Climate change is making such intercession more difficult. As temperatures rise, it takes more rain (or snow) to produce the same amount of runoff. Combined with chronic overuse of the Colorado River, drought and warming have reduced Powell to a puddle of its former self. Since 2000, the lake's surface has dropped by a hundred and forty feet. Just in the past year, it's dropped by fifty feet. As a result, Glen Canyon is beginning to emerge, again, into the light.

THE MORNING AFTER I arrived in Bullfrog, I went back to the marina to meet up with Eric Balken, the executive director of the Glen Canyon Institute. The institute, whose goal is to return the canyon to its natural state, was founded in 1996. A decade later, while Balken was still a student at the University of Utah, he signed on as an intern at the group's office, in Salt Lake City. He's worked there ever since. Now thirty-four, he has probably seen more of Glen Canyon than anyone else under the age of ninety. The first time I spoke to him, over the phone, he offered to show me some "incredible" sights. "It'll be hot," he added.

Again the dock was crowded with families heading out onto Powell in houseboats. For our trip, Balken had rented a pontoon boat. His wife, Sandrine Yang, had decided to come along. So had my husband and two photographers. Once we'd loaded the boat with all our camping gear and supplies, there was only a narrow alley of floor space left.

Balken slipped on a pair of mirrored sunglasses and steered the boat out of the marina, into an arm of the lake known as Bullfrog Bay. From the mouth of the bay, we headed south, into what used to be the main channel of the Colorado. Red cliffs four, five, six hundred feet tall lined the lake on both sides.

As we sped on, the cliffs grew taller and redder. The Colorado used to carry vast amounts of sediment—hence its name, meaning "red-colored." The river, it was said, was "too thick to drink, too thin to plow." Now, though, when the Colorado hits the reservoir's northern edge—a border that keeps creeping south—most of the sediment drops out, leaving the water clear. Lake Powell is an almost tropical shade of turquoise. It sparkled under the cerulean sky. Somewhere deep beneath us, the river was still flowing. But at the surface the water was slack. Yang declared the scene "stupid beautiful."

After about an hour, we arrived at a formation called the Rincon—a mesa with two rocky protuberances sticking out of the top like decaying teeth. Balken had brought along a book with historical photos of Glen Canyon's most famous sites. Pre-dam, the book showed, the Rincon was surrounded by rock and sand. Today, it's lapped by water. A little flat-topped structure bobbed in front of it. A sign announced that this was the Rincon Floating Restroom. I figured it had to be one of the world's most scenic toilets.

From the floating restroom we turned up a side canyon called the Escalante, and from there into a side canyon of the side canyon, Clear Creek. Eventually, the water got so shallow that the boat couldn't go any farther. We jumped out and padded across the wet sand, which was the same ochre as the cliffs.

An enormous rock chamber, known as Cathedral in the Desert, opened before us. The sandstone walls were rounded, but far above our heads they came together, so that the sky was visible only through a slim, S-shaped opening. A curl of sunlight fell on the sand. At the far end of the chamber, a narrow waterfall trickled into a pool. As in a real cathedral, a somber hush prevailed.

In the pre-dam era, Cathedral in the Desert was a sort of natural pilgrimage site—in the words of one visitor, "the end toward which all other wonders had been pointing." In those days, the only way to get there was to hike in, as Clear Creek was too slight to be navigated by boat. Then, for decades, the Cathedral was inaccessible—hidden under Lake Powell—and the waterfall stopped falling, because it, too, was submerged. Balken was thrilled to see the place, if not as it had been—it was still missing any kind of plant life—then at least a lot closer. "This is one of the miracles of Lake Powell being low," he said.

As we were talking, another boat pulled up. Nine people—most of them kids—clambered out. They looked around with dismay. "We're so sad," the oldest woman in the group, who turned out to be the kids' grandmother, told me.

She pointed to a yellow cord, about thirty feet long, hanging from the top of the waterfall. When the lake level had started to drop, someone must have attached it to the rock—I couldn't tell exactly how. It had then been possible to dangle from the cord and leap into the water, and the kids had loved it. Now the rope reached only about halfway down the waterfall and, had there been any way to get up to it, a plunge into the pool would have been fatal.

The woman's husband, who was wearing a Lake Powell cap, seemed to grow suspicious as I took notes. He asked me if I was for or against the reservoir. I tried to duck the question by saying I was a reporter. The group I was with, I acknowledged, was definitely anti.

"I understand the debate," he said. "But the amount of people who enjoy or visit this place versus the number who would if the water wasn't here is astronomical."

"Every time we come to Lake Powell, it's an adventure," the woman said. "But this year it's shockingly disappointing. It's amazing just in one year how much the water has gone down."

"We're on the side that's grateful for the lake," her husband said.

LAKE POWELL IS NAMED for a major in the Union army, John Wesley Powell, who lost an arm in the Battle of Shiloh. Powell served as the second director of the U.S. Geological Survey and organized the first documented ascent of Longs Peak, the highest summit in what's now Rocky Mountain National Park. But he is probably best known for a daring expedition he led down the Green and Colorado Rivers in 1869. In the spring of that year, he set off with nine men and four wooden boats. (One of the boats soon splintered on a rock.) As Powell and his crew explored the still mostly uncharted rivers, they named many of the geological features they encountered—Flaming Gorge, Disaster Falls, Desolation Canyon. A particularly punishing forty-five-mile stretch of river they dubbed Cataract Canyon. After a week spent running Cataract's rapids, Powell and his men were relieved to find themselves drifting on quiet water. In his diary, Powell noted that this more serene stretch of river offered a succession of marvelous sights.

"Past these towering monuments, past these mounded billows of orange sandstone, past these oak-set glens, past these fern-decked alcoves, past these mural curves, we glide hour after hour, stopping now and then, as our attention is arrested by some new wonder," he wrote. He called this stretch Glen Canyon.

Powell spent the next several years leading government-funded surveys of the West. In 1878, he summed up his findings in *Report on the Lands of the Arid Regions of the United States,* delivered to the secretary of the interior. In the report, Powell argued that the West should be regarded almost as a separate country. It should not—indeed, could not—be carved up into hundred-and-sixty-acre plots, the way the Midwest had been; there wasn't enough rain for a farm that size to support a family. Instead, he recommended that parcels be allocated according to a formula that took into account their proximity to water. (Although he was keenly interested in the languages and cultures of Native Americans, Powell seems never even to have questioned the idea that their lands would be handed over to white homesteaders.) He further proposed that Western states be organized according to watershed and that steps be taken to ensure that the region's scarce water resources were shared equitably.

All of his recommendations were ignored. The federal government continued to give out hundred-and-sixty-acre parcels, many of which were obtained fraudulently, by firms or families that amassed vast holdings by gaming the system. Cities began to spring up in places, like the Mojave Desert, with barely enough rainfall to satisfy a scorpion. Clearly, their residents would have to get water from somewhere, and usually that somewhere was the Colorado.

By the nineteen-twenties, there were so many claims on the river that the White House felt compelled to step in. The commerce secretary, Herbert Hoover, was dispatched to Santa Fe to preside over the negotiations, which resulted in the Colorado River Compact. Finalized in the fall of 1922, the compact divided the river at a point in northern Arizona called Lee's Ferry. The states upstream of Lee's Ferry—Colorado, Utah, and Wyoming—along with New Mexico, became known as the upper basin. Collectively, these states were allocated 7.5 million acre-feet of river water a year. Those in the lower basin—California, Arizona, and Nevada—also got 7.5 million acre-feet a year, plus, as a deal sweetener, an extra million. (Later, another 1.5 million acre-feet a year would be promised to Mexico.)

The compact paved—or, if you prefer, lubricated—the way for the creation of the nation's two largest reservoirs, Lake Powell and Lake Mead. Lake Mead sits behind Hoover Dam, completed in 1935, and was designed to serve the lower basin. Today, it supplies practically all the water that's used in Las Vegas and much of what's drunk in cities such as San Diego and Tucson. It also provides—or used to provide, when it was fuller—water for irrigating more than three million acres of corn, cotton, and alfalfa.

Lake Powell, which serves the upper basin, doesn't supply water to much of anyone. Water released from Powell flows into Marble Canyon, then through the Grand Canyon and into Mead. In this sense, Powell is a reservoir for a reservoir. Whether this arrangement ever made sense is unclear. In periods of high flow, Mead should have plenty of water. And in periods of low flow what's the point of impounding the Colorado on its way to Lake Mead?

"You can search and search and search," Matthew Gross, a Utah-based author and political consultant, has written. But, if you want to know why Lake Powell was created, "you'll never find a satisfactory answer." When I asked Jack Schmidt, a professor at Utah State University's Center for Colorado River Studies, to explain the logic behind Powell, he laughed for almost a full minute. "That's a wonderful question," he said, finally.

WHILE I WAS talking to the disappointed family at Cathedral in the Desert, another boat showed up, and then another. Since the lake was crowded and places to camp were few, Balken thought we ought to go claim a spot he had in mind, in a side canyon of the Escalante known as Davis Gulch. When we got there, we found a houseboat with a slide anchored in front.

We continued up the canyon, which twisted in sinuous curves. Along the rock, parallel to the surface of the lake, stretched an unbroken band of white, straight as a ruler. Everywhere you go on Lake Powell, this band is visible. It's made of minerals that the reservoir deposited on the sandstone when it was full and that have been exposed as the water level has dropped. Known as the "bathtub ring," it is now the height of the Statue of Liberty. As the canyon narrowed, I could see that the ring was divided into two distinct layers. The top layer was a brilliant white; the lower layer was speckled with zillions of black dots. The dots, Balken explained, were quagga mussel shells. The mussels, natives of the Black Sea, arrived in Powell about a decade ago, probably on a visitor's boat, and proceeded to multiply madly. The "quagga line," as I came to think of it, shows how far the surface of the lake has fallen since the mussels' numbers exploded.

After a few more turns, we pulled up on a slick of red mud. It was midafternoon and the heat was stultifying. Balken promised "new wonders" to anyone willing to hike up the canyon. Yang declined. The rest of us hopped out. Balken estimated that we were standing on a layer of river sediment twenty or thirty feet deep, all deposited since Lake Powell was created.

The wind had risen, and somehow this only seemed to make it hotter. We wound our way up the canyon, following Davis Gulch's namesake creek. About a half mile from the boat, a huge opening in the cliff face appeared above us. Called La Gorce Arch, it was a window or a porthole in the rock, or, I thought, a blue unblinking eye. Though the canyon was in shadow, the sky, viewed through La Gorce, was radiant. The bathtub ring reached all the way up to the point where, had it been an eye, the lower lashes would be. When Lake Powell was full, Balken said, the arch was a popular destination, and people used to motor right up to it. "In 2019, you could still boat in," he recalled. That afternoon, we had the place to ourselves.

As we hiked on, Balken kept pointing out signs of returning life. "There's a happy willow," he said at one point. "There's a cottonwood," he said at another. Every tadpole we spotted brought an approving murmur. Even a dead beaver, with its buck teeth sticking out of its decomposing skull, seemed to gladden Balken.

"If you were to tally up all these creeks and seeps, it's hundreds of miles of riparian habitat that's coming back," he said. My husband noted that it was a bit awkward to be celebrating the effects of what, by most standards, counts as a disaster.

"I have to admit to a little schadenfreude," Balken said.

In Abbey's novel *The Monkeywrench Gang,* a character called Seldom Seen Smith dubs Lake Powell "the blue death." (The character was modeled on a real river guide named Ken Sleight, who led trips through Glen Canyon in the nineteen-fifties and fought to prevent it from being dammed.) Lake Powell drowned countless creatures outright and killed countless others indirectly, by drowning their food supplies. The death extended well beyond the borders of Glen Canyon. The reservoir changed the flow of the Colorado through the Grand Canyon and so altered its ecosystem, too. Today, several of its native fish species, including the Colorado pikeminnow and the bonytail chub, are drifting toward extinction.

About a mile upstream from La Gorce, we came to a delicate waterfall, maybe forty feet high. It trickled down in stages, as if over a set of stairs. A collection of pointy sticks, plainly the work of beavers, lay at its base. Balken said that when the area had first reappeared from under Lake Powell the waterfall had been buried beneath sediment. At some point, a flash flood must have come crashing down the canyon with enough force to clear it out.

"Clearly, this restoration is happening very quickly," he said cheerfully.

BEFORE IT WAS DROWNED, Glen Canyon was inhabited by humans, off and on, for more than ten thousand years. From an archaeological perspective, its most significant occupants were the people known as the Ancestral Puebloans, or, in Navajo, the Anasazi.

At the height of their influence, the Ancestral Puebloans controlled a vast swath of the Four Corners region. Their settlements included Chaco Canyon, where the tallest pre-skyscraper buildings in North America were erected, in the tenth and eleventh centuries, and Mesa Verde, where a fantastic cliff city arose in the twelfth century. To the Puebloans, Glen Canyon was farm country. They grew maize, squash, and cotton on the floors of its side canyons and impounded its streams and seeps to store water for their crops. So that they could travel overland between one side canyon and the next, they chiseled footholds into the cliff faces.

Starting around the year 1200, Glen Canyon experienced a population boom. Then, just sixty or seventy years later, the place emptied out. The granaries, the kivas, and the stone cliff dwellings were abandoned. And what held for Glen Canyon held for virtually all of the other settlements in the area.

The cause of the collapse remains a mystery. One theory blames the weather. Tree-ring records have allowed researchers to reconstruct soil moisture in the region, year by year, going back to the ninth century. What's sometimes called the "great drought" began in the twelve-seventies and lasted through the twelve-nineties. By the time the streams started to run again, no one was left to make use of them.

According to a recent paper in *Science,* the drought that's plagued the Southwest since the early two-thousands is already more acute than the worst stretch of the great drought. It's also worse than an even greater drought that hit the region in the mid-eleven-hundreds and nearly as bad as the most severe dry spell in the record, which occurred in the late fifteen-hundreds. Indeed, the authors of the paper concluded, all of western North America, which includes northern Mexico, is currently on a "megadrought-like trajectory." Withdrawals from Lake Powell and Lake Mead have helped mask the severity of the situation, but what happens next?

Park Williams, a climate scientist at UCLA and the lead author of the *Science* study, told me that, to researchers, the droughts indicated in the tree-ring record "were almost like mythical beasts, lurking there." Those droughts, it's believed, were caused by shifts in the temperature of the eastern Pacific, which produced air circulation patterns that blocked storms from reaching the western part of the continent. Today, too, naturally occurring oscillations in sea surface temperatures are keeping the West dry. But now there's also climate change to contend with.

"The only way to get an exceptional event is to have bad luck," Williams said. "And the bad luck comes from the tropical Pacific Ocean. But this event isn't only bad luck. It's also a very straightforward effect of global warming."

He went on, "Warmer air evaporates water out of soils and ecosystems more quickly. So every raindrop or snowflake becomes a bit less potent, because the atmosphere has this increasing thirst. And that means that as we go into the future, to get into a drought as bad as the one we're in now, it's going to take less and less bad luck, because human-caused warming is doing more and more of the heavy lifting."

According to another study, published in the journal *Water Resources Research,* during the first fourteen years of the twenty-first century the average flow of the Colorado River was almost 20 percent lower than it was during the twentieth century. The authors of this study, Brad Udall, of Colorado State University, and Jonathan Overpeck, of the University of Michigan, attributed a third of the decline to global warming. They predicted that, as temperatures continue to rise, the amount of water in the river will continue to drop. "It is imperative that decision-makers begin to consider seriously the policy implications of potential large-scale future flow declines," they warned.

"The world is rife with examples where it looks like drought played a role—maybe not *the* role—in destabilizing societies," Overpeck, who is a paleoclimatologist, told me. "It seems like that happened a lot."

AS LAKE POWELL RECEDES, Glen Canyon's archaeological sites are gradually resurfacing. Our second morning on the lake, Balken decided that we should go look for one. A bass fisherman had told him he'd seen the ruins of a stone building in an alcove near the entrance to the Escalante. (Bass were introduced into Lake Powell in the nineteen-sixties and, along with quagga mussels, have pretty much taken over the ecosystem.)

What exactly the fisherman meant, Balken wasn't sure. We spent a few hours zigging and zagging along the cliffs near the mouth of the Escalante, trying to peer into whatever opening might qualify as an "alcove." The weather had changed; instead of very hot and clear, it was now very hot and cloudy. Against the gray, the amber cliffs seemed more sphinx-like and more menacing.

We checked out more than a dozen alcove-like hollows but never did manage to locate the ruins. The one trace of the Ancestral Puebloans that Balken spotted was a line of footholds snaking up an almost vertical rock face. Trying to imagine someone using the footholds was, I found, vertigo-inducing.

That afternoon, we motored up another side canyon of the Colorado—Iceberg Canyon. Around every curve, there was a houseboat sitting at anchor; in one spot, two houseboats were parked side by side, surrounded by a small beach's worth of inflatable toys. After a few miles, we once again hit mud. In front of us stood a grove of dead cottonwoods. Several of the trees were festooned with plastic jugs. Balken explained that the empty jugs had served as buoys when the trees first started to reemerge from the lake, presumably to prevent boats from getting snagged on them. The jugs now dangled twenty-five or thirty feet off the ground. As we disembarked, an osprey took off from one of the cottonwoods' silvery branches.

We started walking across what seemed like a Sahara of red sand. "When I look at this canyon, I think, 'There's a lot of sediment to be moved here,'" Balken said.

The hike led uphill, and the farther we walked the more vegetation there was, until we found ourselves bushwhacking through thick shrubbery. Moonflowers bloomed, ghostly pale against their dark foliage. A canyon wren sang, soulfully, somewhere in the distance. Balken pointed out that we'd climbed above the bathtub ring, which meant that, for the first time on the trip, we'd reached a part of Glen Canyon that had never been flooded. He sniffed the air. "It smells alive," he observed.

Soon, we came to a stone amphitheater. Fifty feet up, a ring of greenery clung to the walls—a hanging garden. We sat for a while, admiring the garden and enjoying the amphitheater's damp shade. I recalled a story I'd read about Barry Goldwater, a man not generally known as an environmentalist. Before he launched his political career, Goldwater took a trip down the Colorado that was supposed to re-create John Wesley Powell's famous journey. When he finally retired, after five terms representing Arizona in the U.S. Senate and one failed presidential bid, Goldwater said that the only vote he regretted having cast was the one that led to the damming of Glen Canyon.

"I think of that river as it was when I was a boy," he said. "And that is the way I would like to see it again."

GLEN CANYON DAM was approved by Congress in the spring of 1956, as part of an extensive infrastructure bill that also authorized the construction of Flaming Gorge Dam, on the Green River, Navajo Dam, on the San Juan, and Blue Mesa Dam, on the Gunnison. According to a history published by the Bureau of Reclamation, Glen Canyon Dam was designed to serve as a "cash register" that would cover the cost of the other, smaller projects by producing hydropower, which the bureau would sell to utility companies. If the dam is hard to explain as a water-management tool, that may be because it wasn't intended as one.

Before work on Glen Canyon Dam could begin in earnest, the Colorado River had to be channeled out of the way. Contractors blasted two enormous diversion tunnels into the sandstone near Page, Arizona, a town built from scratch to house the project's workforce. When the dam was completed, in 1963, the tunnels were sealed off with reinforced concrete. Today, water exits Lake Powell through eight pipes, or penstocks, equipped with turbines. If the current drought continues, then within a couple of years the surface of the reservoir could fall below what's known as "power pool." At that point, water would no longer flow through the penstocks, so the dam would no longer produce electricity or, by extension, revenue. Already, the Bureau of Reclamation is concerned enough about this possibility that it's releasing water from upstream reservoirs, like Flaming Gorge, to try to boost Lake Powell's level.

"We hoped to never go down this road," Wayne Pullan, the director of the bureau's Upper Colorado River Region, said in announcing the move in July. "But now we have to." As the writer Rebecca Solnit noted after visiting Lake Powell a few years ago, "The future we foresee is often not the one we get."

Today, the Bureau of Reclamation operates Lake Mead and Lake Powell so as to keep, roughly speaking, the same amount of water in each. As a result, both reservoirs are now at only about a third of their capacity, meaning there's not enough water to fill even one of them.

To Balken and his colleagues at the Glen Canyon Institute, this is a crisis that shouldn't go to waste. Under a proposal that the institute calls Fill Mead First, water from the Colorado, instead of being divided between the two reservoirs, would be sent straight on to Mead. Powell would then contract until most—perhaps even all—of Glen Canyon resurfaced.

With "total storage between Powell and Mead reaching record lows, the proposal to Fill Mead First becomes more realistic and pragmatic every day," the Glen Canyon Institute's website argues.

"If Powell goes down another forty feet, that's just spitting distance from power pool," Balken told me. "And as soon as that threshold gets passed, all the incentives change. The whole conversation changes. I think people just don't want to admit it to themselves."

But the obstacles to filling Mead first are practically as large as the reservoirs themselves. If the level of Lake Powell keeps dropping, it becomes harder and harder to get water out, until the reservoir reaches what's known as "dead pool." In its most ambitious form, Fill Mead First would entail drilling new bypass tunnels around Glen Canyon Dam so that the Colorado could run—sort of—as it used to. No one has made a serious estimate of what this would cost.

Jack Schmidt, the Utah State University professor, has studied the Fill Mead First proposal and also, as an intellectual exercise, the option of doing the reverse—filling Powell and letting Mead empty. Neither of these proposals, he's concluded, does much to solve the basic problem, which is that there's not enough water to fulfill the terms of the Colorado River Compact—and there probably never was.

"I personally am not going to tell you whether Fill Mead First or Fill Powell First or equalization is a better idea, because I'm unfortunately too aware of the complications of each of them," Schmidt told me. "They're just every one of them a no-win situation."

Anne Castle is a senior fellow at the University of Colorado Law School who served as the assistant secretary for water and science at the Department of the Interior in the Obama administration. (The Bureau of Reclamation is a division of Interior.) "Big sectors of the economy have grown up in reliance on those two reservoirs operating in the way they do now," Castle told me. One of these sectors is recreation. Though Easterners have barely heard of it, Lake Powell is one of the National Park Service's most popular attractions. In a good year, it draws more than four million visitors, who collectively spend almost half a billion dollars.

"We're facing big challenges, and so I think that radical ideas need to be on the table and be examined," Castle continued. "The main problem, though, is we're using too much water, no matter where you put it."

ON MY WAY BACK to Salt Lake City, I decided to stop in Moab. A huge column of smoke was rising from the mountains south of town, where an abandoned campfire had burgeoned into a nine-thousand-acre forest fire. Many of the houses sported hand-painted signs that said "THANK YOU FIREFIGHTERS!"

Ken Sleight, the model for Abbey's character Seldom Seen Smith, is now ninety-one and lives on a farm not far from Moab. I'd hoped to go talk to him about his memories of Glen Canyon prior to the "blue death." But, a few days before I arrived, the blaze in the mountains had swept through the farm, destroying a building that Sleight had used to store records of, among many other things, his long career as a river runner and the fight against the dam. Through a friend, he let it be known that he didn't feel up to an interview.

Instead, I went to talk to Mike DeHoff, the founder of a project called Returning Rapids. DeHoff owns a business in Moab that fabricates aluminum frames for the sort of rafts used to run the Colorado. When I got to his workshop, welding had ceased for the day and the place was quiet, but DeHoff still had a pair of earplugs dangling on a string around his neck. The first thing he pointed out to me was a dinghy hanging on the wall. Made of molded plastic, it was about eight feet long and the color of rhubarb. At least according to legend, it had belonged to Abbey, who lived in Moab in the late nineteen-seventies. "He passed it on to somebody, and they said it needed to be someplace like this," DeHoff told me.

Before he opened his welding business, DeHoff worked as a professional guide on the Colorado, and he's still an avid river runner. His favorite stretch of water is Cataract Canyon, just upstream from Glen Canyon. To a lesser but still significant extent, Cataract, too, was flooded by Lake Powell. When John Wesley Powell traveled it, Cataract presented him and his men with fifty sets of rapids, some so intimidating that they had to portage their heavy wooden boats around them. When Lake Powell was full, the water backed up far enough into Cataract that more than half the rapids disappeared.

"It used to be that you would go through the twenty-third rapid in Cataract Canyon, which is known as Big Drop 3 or Satan's Gut, and you'd see houseboats," DeHoff recalled. "It was such a contrast, because you were on this wild river, and then, boom, you'd watch the river die." Then, around 2005, DeHoff began to notice that some of the drowned rapids were returning. As Powell continued to shrink, more rapids re-emerged. Slowed by the reservoir, the river had dumped an enormous amount of sediment in the canyon; DeHoff dubbed the resulting mud-flats the Dominy Formation, for Floyd Dominy, who served as the commissioner of the Bureau of Reclamation in the nineteen-sixties and was Glen Canyon Dam's biggest booster. The river was cutting new channels through the sediment, with unpredictable results; from year to year, and even month to month, it was hard to know what to expect. DeHoff put together a spreadsheet of the "returning rapids," which he kept updating. His wife, Meg Flynn, a librarian, began collecting historical photographs of Cataract, to use for comparison. Eventually, several scientists became involved in the project: here was geology happening in real time.

"We're seeing a lot of the river trying to restore itself," DeHoff told me. "And it's been fascinating to watch."

After a while, Flynn showed up at the shop, along with Peter Lefebvre, a professional river guide, and Chris Benson, a geologist turned pilot. Lefebvre had just led a raft trip through Cataract Canyon; DeHoff and Flynn were setting out on one the following day. The conversation turned to the difficulty of getting rafts out of the water at the end of a trip. The National Park Service had built a concrete boat ramp for this purpose, but now, thanks to Lake Powell's contraction, it had been left high and dry. Various fixes had been attempted, but the water was receding so fast that these kept failing. Further complicating the situation, the Colorado, instead of following its historical course, was gouging a new channel in the area. To haul out their boats, Lefebvre reported, some groups were resorting to hooking up two pickup trucks, one behind the other.

Flynn shook her head. "Someone could get killed," she said.

I tried to steer the conversation to the future of Lake Powell. What should be done?

"I think the Fill Mead First proposal—there's a lot of merit to that," DeHoff said. "Are people going to be happy with it? No. Are people going to be happy with any solution you come up with? I don't think so."

"I don't know how many people go to Lake Powell, but it's probably an order of magnitude greater than the number of people who could run the river," Benson said. "A lot of people love that lake. So we'd have to find the right way to make everybody equally unhappy."

"I feel like Mother Nature's kind of forcing the hand of people to make decisions that are really hard decisions," Lefebvre said.

Moab sits next to Arches National Park, where most of Abbey's

Desert Solitaire is set. (Before moving to the town, Abbey spent three seasons as a ranger in the park.) He describes Arches as an "inhuman spectacle of rock and cloud and sky and space." But he warns his readers against going to visit it; the place has been destroyed by, among other things, too many tourists.

"This is not a travel guide but an elegy," he writes. I decided to ignore him. The next day, I got up early and hiked out to one of the park's most famous spots, Delicate Arch, by seven a.m. So many visitors had got there ahead of me that there was a long line of people waiting to take a selfie with the arch. I was glad I had come, because it was such a remarkable sight.

Originally published in *The New Yorker,*
August 16, 2021.

Update: In early 2023, the water level in Lake Powell dropped to an all-time low and the reservoir was filled to only 22 percent of capacity. Since then, the water level has risen, but it remains low by historical standards and the long-term forecast is grim. Eric Balken continues to press for the restoration of Glen Canyon.

THE ISLAND IN THE WIND

A Danish Community Goes Carbon Neutral

JØRGEN TRANBERG IS a farmer who lives on the Danish island of Samsø. He is a beefy man with a mop of brown hair and an unpredictable sense of humor. When I arrived at his house one gray morning this spring, he was sitting in his kitchen, smoking a cigarette and watching grainy images on a black-and-white TV. The images turned out to be closed-circuit shots from his barn. One of his cows, he told me, was about to give birth, and he was keeping an eye on her. We talked for a few minutes, and then, laughing, he asked me if I wanted to climb his wind turbine. I was pretty sure I didn't, but I said yes anyway.

We got into Tranberg's car and bounced along a rutted dirt road. The turbine loomed up in front of us. When we reached it, Tranberg stubbed out his cigarette and opened a small door in the base of the tower. Inside were eight ladders, each about twenty feet tall, attached one above the other. We started up, and were soon huffing. Above the last ladder, there was a trapdoor, which led to a sort of engine room. We scrambled into it, at which point we were standing on top of the generator. Tranberg pressed a button, and the roof slid open to reveal the gray sky and a patchwork of green and brown fields stretching toward the sea. He pressed another button. The rotors, which he had switched off during our climb, started to turn, at first sluggishly and then much more rapidly. It felt as if we were about to take off. I'd like to say the feeling was exhilarating, but in fact I found it sickening. Tranberg looked at me and started to laugh.

Samsø, which is roughly the size of Nantucket, sits in what's known as the Kattegat, an arm of the North Sea. The island is bulgy in the south and narrows to a bladelike point in the north, so that on a map it looks a bit like a woman's torso and a bit like a meat cleaver. It has twenty-two villages that hug the narrow streets; out back are fields where farmers grow potatoes and wheat and strawberries. Thanks to Denmark's peculiar geography, Samsø is smack in the center of the country and, at the same time, in the middle of nowhere.

For the past decade or so, Samsø has been the site of an unlikely social movement. When it began, in the late nineteen-nineties, the island's forty-three hundred inhabitants had what might be described as a conventional attitude toward energy: as long as it continued to arrive, they weren't much interested in it. Most Samsingers heated their houses with oil, which was brought in on tankers. They used electricity imported from the mainland via cable, much of which was generated by burning coal. As a result, each Samsinger put into the atmosphere, on average, nearly eleven tons of carbon dioxide annually.

Then, quite deliberately, the residents of the island set about changing this. They formed energy cooperatives and organized seminars on wind power. They removed their furnaces and replaced them with heat pumps. By 2001, fossil-fuel use on Samsø had been cut in half. By 2003, instead of importing electricity, the island was exporting it, and by 2005 it was producing from renewable sources more energy than it was using.

The residents of Samsø that I spoke to were clearly proud of their accomplishment. All the same, they insisted on their ordinariness. They were, they noted, not wealthy, nor were they especially well educated or idealistic. They weren't even terribly adventuresome. "We are a conservative farming community" is how one Samsinger put it.

"We are only normal people," Tranberg told me. "We are not some special people."

THIS YEAR, the world is expected to burn through some thirty-one billion barrels of oil, six billion tons of coal, and a hundred trillion cubic feet of natural gas. The combustion of these fossil fuels will produce, in aggregate, some four hundred quadrillion BTUs of energy. It will also yield around thirty billion tons of carbon dioxide. Next year, global consumption of fossil fuels is expected to grow by about 2 percent, meaning that emissions will rise by more than half a billion tons, and the following year consumption is expected to grow by yet another 2 percent.

When carbon dioxide is released into the air, about a third ends up, in relatively short order, in the oceans. (CO_2 dissolves in water to form a weak acid; this is the cause of the phenomenon known as "ocean acidification.") A quarter is absorbed by terrestrial ecosystems—no one is quite sure exactly how or where—and the rest remains in the atmosphere. If current trends in emissions continue, then sometime within the next four or five decades the chemistry of the oceans will have been altered to such a degree that many marine organisms—including reef-building corals—will be pushed toward extinction. Meanwhile, atmospheric CO_2 levels are projected to reach five hundred and fifty parts per million—twice preindustrial levels—virtually guaranteeing an eventual global temperature increase of three or more degrees. The consequences of this warming are difficult to predict in detail, but even broad, conservative estimates are terrifying: at least 15 and possibly as many as 30 percent of the planet's plant and animal species will be threatened; sea levels will rise by several feet; yields of crops like wheat and corn will decline significantly in a number of areas where they are now grown as staples; regions that depend on glacial runoff or seasonal snowmelt—currently home to more than a billion people—will face severe water shortages; and what now counts as a hundred-year drought will occur in some parts of the world as frequently as once a decade.

Today, with CO_2 levels at three hundred and eighty-five parts per million, the disruptive impacts of climate change are already apparent. The Arctic ice cap, which has shrunk by half since the nineteen-fifties, is melting at an annual rate of twenty-four thousand square miles, meaning that an expanse of ice the size of West Virginia is disappearing each year. Over the past ten years, forests covering a hundred and fifty million acres in the United States and Canada have died from warming-related beetle infestations. It is believed that rising temperatures are contributing to the growing number of international refugees—"Climate change is today one of the main drivers of forced displacement," the United Nations' high commissioner for refugees, António Guterres, said recently—and to armed conflict: some experts see a link between the fighting in Darfur, which has claimed as many as three hundred thousand lives, and changes in rainfall patterns in equatorial Africa.

"If we keep going down this path, the Darfur crisis will be only one crisis among dozens of others," President Nicolas Sarkozy, of France, told a meeting of world leaders in April. The secretary-general of the United Nations, Ban Ki-moon, has called climate change "the defining challenge of our age."

In the context of this challenge, Samsø's accomplishments could be seen as trivial. Certainly, in numerical terms they don't amount to much: all the island's avoided emissions of the past ten years are overwhelmed by the CO_2 that a single coal-fired power plant will emit in the next three weeks, and China is building new coal-fired plants at the rate of roughly four a month. But it is also in this context that the island's efforts are most significant. Samsø transformed its energy systems in a single decade. Its experience suggests how the carbon problem, as huge as it is, could be dealt with, if we were willing to try.

SAMSØ SET OUT to reinvent itself thanks to a series of decisions that it had relatively little to do with. The first was made by the Danish Ministry of Environment and Energy in 1997. The ministry, looking for ways to promote innovation, decided to sponsor a renewable-energy contest. In order to enter, a community had to submit a plan showing how it could wean itself off fossil fuels. An engineer who didn't actually live on Samsø thought the island would make a good candidate. In consultation with Samsø's mayor, he drew up a plan and submitted it. When it was announced that Samsø had won, the general reaction among residents was puzzlement. "I had to listen twice before I believed it," one farmer told me.

The brief surge of interest that followed the announcement soon dissipated. Besides its designation as Denmark's "renewable-energy island," Samsø received basically nothing—no prize money or special tax breaks, or even government assistance. One of the few people on the island to think the project was worth pursuing was Søren Hermansen.

Hermansen, who is now forty-nine, is a trim man with close-cropped hair, ruddy cheeks, and dark-blue eyes. He was born on Samsø and, save for a few stints away to travel and go to university, has lived there his entire life. His father was a farmer who grew, among other things, beets and parsley. Hermansen, too, tried his hand at farming—he took over the family's hundred acres when his father retired—but he discovered he wasn't suited to it.

"I like to talk, and vegetables don't respond," he told me. He leased his fields to a neighbor and got a job teaching environmental studies at a local boarding school. Hermansen found the renewable-energy-island concept intriguing. When some federal money was found to fund a single staff position, he became the project's first employee.

For months, which stretched into years, not much happened. "There was this conservative hesitating, waiting for the neighbor to do the move," Hermansen recalled. "I know the community and I know this is what usually happens." Rather than working against the islanders' tendency to look to one another, Hermansen tried to work with it.

"One reason to live here can be social relations," he said. "This renewable-energy project could be a new kind of social relation, and we used that." Whenever there was a meeting to discuss a local issue—any local issue—Hermansen attended and made his pitch. He asked Samsingers to think about what it would be like to work together on something they could all be proud of. Occasionally, he brought free beer along to the discussions. Meanwhile, he began trying to enlist the support of the island's opinion leaders. "This is where the hard work starts, convincing the first movers to be active," he said. Eventually, much as Hermansen had hoped, the social dynamic that had stalled the project began to work in its favor. As more people got involved, others were prompted to do so. After a while, enough Samsingers were participating that participation became the norm.

"People on Samsø started thinking about energy," Ingvar Jørgensen, a farmer who heats his house with solar hot water and a straw-burning furnace, told me. "It became a kind of sport."

"It's exciting to be a part of this," Brian Kjær, an electrician who installed a small-scale turbine in his backyard, said. Kjær's turbine, which is seventy-two feet tall, generates more current than his family of three can use, and also more than the power lines leading away from his house can handle, so he uses the excess to heat water, which he stores in a tank that he rigged up in his garage. He told me that one day he would like to use the leftover electricity to produce hydrogen, which could potentially run a fuel-cell car.

"Søren, he has talked again and again, and slowly it's spread to a lot of people," he said.

SINCE BECOMING THE "renewable-energy island," Samsø has increasingly found itself an object of study. Researchers often travel great distances to get there, a fact that is not without its own irony. The day after I arrived, from New York via Copenhagen, a group of professors from the University of Toyama, in Japan, came to look around. They had arranged a tour with Hermansen, and he invited me to tag along. We headed off to meet the group in his electric Citroen, which is painted blue with white puffy clouds on the doors. It was a drizzly day, and when we got to the dock the water was choppy. Hermansen commiserated with the Japanese, who had just disembarked from the swaying ferry; then we all boarded a bus.

Our first stop was a hillside with a panoramic view of the island. Several wind turbines exactly like the one I had climbed with Tranberg were whooshing nearby. In the wet and the gray, they were the only things stirring. Off in the distance, the silent fields gave way to the Kattegat, where another group of turbines could be seen, arranged in a soldierly line in the water.

All told, Samsø has eleven large land-based turbines. (It has about a dozen additional micro-turbines.) This is a lot of turbines for a relatively small number of people, and the ratio is critical to Samsø's success, as is the fact that the wind off the Kattegat blows pretty much continuously; flags on Samsø, I noticed, do not wave—they stick straight out, as in children's drawings. Hermansen told us that the land-based turbines are a hundred and fifty feet tall, with rotors that are eighty feet long. Together, they produce some twenty-six million kilowatt-hours a year, which is just about enough to meet all the island's demands for electricity. (This is true in an arithmetic sense; as a practical matter, Samsø's production of electricity and its needs fluctuate, so that sometimes it is feeding power into the grid and sometimes it is drawing power from it.) The offshore turbines, meanwhile, are even taller—a hundred and ninety-five feet high, with rotors that extend a hundred and twenty feet. A single offshore turbine generates roughly eight million kilowatt-hours of electricity a year, which, at Danish rates of energy use, is enough to satisfy the needs of some two thousand homes. The offshore turbines—there are ten of them—were erected to compensate for Samsø's continuing use of fossil fuels in its cars, trucks, and ferries. Their combined output, of around eighty million kilowatt-hours a year, provides the energy equivalent of all the gasoline and diesel oil consumed on the island and then some; in aggregate, Samsø generates about 10 percent more power than it consumes.

"When we started, in 1997, nobody expected this to happen," Hermansen told the group. "When we talked to local people, they said, Yes, come on, maybe in your dreams." Each land-based turbine cost the equivalent of eight hundred and fifty thousand dollars. Each offshore turbine cost around three million dollars. Some of Samsø's turbines were erected by a single investor, like Tranberg; others were purchased collectively. At least four hundred and fifty island residents own shares in the onshore turbines, and a roughly equal number own shares in those offshore. Shareholders, who also include many nonresidents, receive annual dividend checks based on the prevailing price of electricity and how much their turbine has generated.

"If I'm reduced to being a customer, then if I like something I buy it, and if I don't like it I don't buy it," Hermansen said. "But I don't care about the production. We care about the production, because we own the wind turbines. Every time they turn around, it means money in the bank. And, being part of it, we also feel responsible." Thanks to a policy put in place by Denmark's government in the late nineteen-nineties, utilities are required to offer ten-year fixed-rate contracts for wind power that they can sell to customers elsewhere. Under the terms of these contracts, a turbine should—barring mishap—repay a shareholder's initial investment in about eight years.

From the hillside, we headed to the town of Ballen. There we stopped at a red shed-shaped building made out of corrugated metal. Inside, enormous bales of straw were stacked against the walls. Hermansen explained that the building was a district heating plant that had been designed to run on biomass. The bales, each representing the equivalent of fifty gallons of oil, would be fed into a furnace, where water would be heated to a hundred and fifty-eight degrees. This hot water would then be piped underground to two hundred and sixty houses in Ballen and in the neighboring town of Brundby. In this way, the energy of the straw burned at the plant would be transferred to the homes, where it could be used to provide heat and hot water.

Samsø has two other district heating plants that burn straw—one in Tranebjerg, the other in Onsbjerg—and also a district plant, in Nordby, that burns wood chips. When we visited the Nordby plant, later that afternoon, it was filled with what looked like mulch. (The place smelled like a potting shed.) Out back was a field covered in rows of solar panels, which provide additional hot water when the sun is shining. Between the rows, sheep with long black faces were munching on the grass. The Japanese researchers pulled out their cameras as the sheep snuffled toward them expectantly.

Of course, burning straw or wood, like burning fossil fuels, produces CO_2. The key distinction is that while fossil fuels release carbon that otherwise would have remained sequestered, biomass releases carbon that would have entered the atmosphere anyway, through decomposition. As long as biomass regrows, the CO_2 released in its combustion should be reabsorbed, meaning that the cycle is—or at least can be—carbon neutral. The wood chips used in the Nordby plant come from fallen trees that previously would have been left to rot. The straw for the Ballen-Brundby plant comes mainly from wheat stalks that would previously have been burned in the fields. Together, the biomass heating plants prevent the release of some twenty-seven hundred tons of carbon dioxide a year.

In addition to biomass, Samsø is experimenting on a modest scale with biofuels: a handful of farmers have converted their cars and tractors to run on canola oil. We stopped to visit one such farmer, who grows his own seeds, presses his own oil, and feeds the leftover mash to his cows. The farmer couldn't be located, so Hermansen started up the press himself. He stuck a finger under the spout, then popped it into his mouth. "The oil is very good," he announced. "You can use it in your car, and you can use it on your salad."

After the tour, I went back with Hermansen to his office, in a building known as the Energiakademi. The academy, which looks like a Bauhaus interpretation of a barn, is covered with photovoltaic cells and insulated with shredded newspapers. It is supposed to serve as a sort of interpretive center, though when I visited, the place was so new that the rooms were mostly empty. Some high-school students were kneeling on the floor, trying to put together a miniature turbine.

I asked Hermansen whether there were any projects that hadn't worked out. He listed several, including a plan to use natural gas produced from cow manure and an experiment with electric cars that failed when one of the demonstration vehicles spent most of the year in the shop. The biggest disappointment, though, had to do with consumption.

"We made several programs for energy savings," he told me. "But people are acting—what do you call it?—irresponsibly. They behave like monkeys." For example, families that insulated their homes better also tended to heat more rooms, "so we ended up with zero." Essentially, he said, energy use on the island has remained constant for the past decade.

I asked why he thought the renewable-energy-island effort had got as far as it did. He said he wasn't sure, because different people had had different motives for participating. "From the very egoistic to the more over-all perspective, I think we had all kinds of reasons."

Finally, I asked what he thought other communities might take from Samsø's experience.

"We always hear that we should think globally and act locally," he said. "I understand what that means—I think we as a nation should be part of the global consciousness. But each individual cannot be part of that. So 'Think locally, act locally' is the key message for us."

"There's this wish for showcases," he added. "When we are selected to be the showcase for Denmark, I feel ashamed that Denmark doesn't produce anything bigger than that. But I feel proud because we are the showcase. So I did my job, and my colleagues did their job, and so did the people of Samsø."

AROUND THE SAME TIME that Samsø was designated Denmark's renewable-energy island, a group of Swiss scientists who were working on similar issues performed a thought experiment. The scientists, all of whom were affiliated with the Swiss Federal Institute of Technology, asked themselves what level of energy use would be sustainable, not just for an island or a small European nation, but for the entire world. The answer they came up with—two thousand watts per person—furnished the name for a new project: the 2,000-Watt Society.

"What it's important, I think, to know is that the 2,000-Watt Society is not a program of hard life," the director of the project, Roland Stulz, told me when I went to speak to him at his office, in the Zurich suburb of Dübendorf. "It is not what we call *Gürtel enger schnallen*"—belt tightening—"it's not starving, it's not having less comfort or fun. It's a creative approach to the future."

Stulz, who is sixty-three, is a soft-spoken man with dark wavy hair and a salt-and-pepper mustache. He was trained as an architect and later became interested in energy-efficient building. In 2001, when he took over the 2,000-Watt Society, his mandate was to push it into the realm of the practical. (His work is funded in part by the Swiss Federal Institute of Technology, which has campuses in Zurich and Lausanne, and in part by private donations.) He began holding meetings that brought researchers together with government officials from cities like Zurich and Basel.

"I divided them into groups," Stulz recalled. "And I told them, 'At four o'clock each group must come and tell the whole session what project they will do in the future, and who will lead the projects.' And they said, 'Oh, it's not possible.' But at four o'clock everybody came with a project. And that's how we started." The cantons of Geneva and Basel-Stadt and the city of Zurich subsequently endorsed the aims of the 2,000-Watt Society, as did the Swiss Federal Department of the Environment, Transport, Energy, and Communications. "At first glance, the objective of a two-thousand-watt society appears unrealistic," Moritz Leuenberger, the head of the federal department, has said. "But the necessary technology already exists."

One afternoon, Stulz took me to visit the headquarters of an aquatic-research center known as EAWAG, which was designed to meet the 2,000-Watt Society's energy efficiency goals. (EAWAG is an acronym for a German name so complicated that even German speakers can't remember it.) We drove over in his Volvo, which runs on compressed natural gas produced in part from rotting vegetables. When I first caught sight of the place, I thought it was covered with banners; these turned out to be tinted-glass panels. Inside, hanging from a set of chains in a large atrium, was what I took to be a model of a water molecule, enlarged some ten billion times.

Among the many unusual features of the EAWAG Center is a lack of usual features. The building, which opened in 2006, has no furnace; it is so tightly insulated that, on most days, the warmth thrown off by the office equipment and the two hundred people who work inside is enough to keep it comfortable. Additional heat is provided by the sun—in winter, the outside panels tilt to allow in the maximum amount of light—and by air sucked in from underground. The building also has no conventional air conditioners: in summer, the panels tilt to provide shade, and if the building gets hot during the day, at night the windows at the top of the atrium open, and the warm air rushes out. It supplies about a third of its own electricity with photovoltaic panels installed on the roof and gets its hot water from solar collectors. Its bathrooms are equipped with specially designed "no mix" toilets that separate out urine, which contains potentially useful phosphorus and nitrogen. ("Exploiting common waste as a resource is a mark of sustainable civilization," a booklet on the building observes.)

"It's not a miracle, such a building," Stulz told me when we went to have a cup of coffee in the center's cheerfully modernist cafeteria. "It's just putting smart elements together in a smart way." Outside, it was rainy and forty-three degrees; inside the temperature was a pleasant seventy.

ONE WAY TO think about the 2,000-Watt Society is in terms of light bulbs. Let's say you turn on twenty lamps, each with a hundred-watt bulb. Together, the lamps will draw two thousand watts of power. Left on for a day, they will consume forty-eight kilowatt-hours of energy; left on for a year, they will consume seventeen thousand five hundred and twenty kilowatt-hours. A person living a two-thousand-watt life would consume in all his activities—working, eating, traveling—the same amount of energy as those twenty bulbs, or seventeen thousand five hundred and twenty kilowatt-hours annually.

Most of the people in the world today consume far less than this. The average Bangladeshi, for example, uses only about twenty-six hundred kilowatt-hours a year—this figure includes all forms of energy, from electricity to transportation fuel—which is the equivalent of using roughly three hundred watts continuously. The average Indian uses about eighty-seven hundred kilowatt-hours a year, making India a one-thousand-watt society, while the average Chinese uses about thirteen thousand kilowatt-hours a year, making China a fifteen-hundred-watt society.

Those of us who live in the industrialized world, by contrast, consume far more than two thousand watts. Switzerland, for instance, is a five-thousand-watt society. Most other western European countries are six-thousand-watt societies; the United States and Canada run at twelve thousand watts. One of the founding principles of the

2,000-Watt Society is that this disparity is in itself unsustainable. "It's a basic matter of fairness" is how Stulz put it to me. But increasing energy use in developing countries to match that of industrialized nations would be unacceptable on ecological grounds. Were per capita demand in the developing world to reach current European levels, global energy consumption would more than double, and were it to rise to the American level, global energy consumption would more than triple. The 2,000-Watt Society gives industrialized countries a target for cutting energy use at the same time that it sets a limit for growth in developing nations.

The last time Switzerland was a two-thousand-watt society was in the early nineteen-sixties. By the end of that decade, energy use had reached three thousand watts, and by the mid-seventies it was up to four thousand watts. This rapid rise could be said to follow from technological advances—the spread of automobiles, the advent of jet travel, the proliferation of appliances and electronic devices—or it could be seen as just the reverse: a failure to apply technology where it was needed. A few years ago, a group of Swiss scientists published a white paper—or, to use the Swiss term, a "white book"—on the feasibility of a two-thousand-watt society. Relying on widely agreed-upon figures, the scientists estimated that two-thirds of all the primary energy consumed in the world today is wasted, mostly in the form of heat that nobody wants or uses. ("Primary energy" is the energy contained in, say, a lump of coal; "useful energy" is the light emitted by a bulb once that coal has been burned to produce steam, the steam has been used to run a turbine, and the resulting electricity has been transmitted over the grid to heat the bulb's filament.) This same paper concluded that, with currently available technologies, buildings could be made 80 percent more efficient, cars 50 percent more efficient, and industrial motors 25 percent more efficient.

In Switzerland, I visited several other buildings that, like the EAWAG Center, had been specifically designed to maximize efficiency. One was an upscale apartment building in Basel. The apartments have eighteen-inch-thick walls filled with insulation, triple-paned windows coated with a special reflective film, and a heat-recovery system that captures 80 percent of the energy normally lost through ventilation. Instead of a boiler, it has a geothermal heat pump, which essentially sucks energy out of the groundwater. In the summer, the same system is used for cooling. (In compliance with Swiss building codes, the building also contains a bomb shelter.)

"The construction industry is very traditional," Franco Fregnan, an engineer who showed me around the apartments, said. "If you bring an innovation to them, you usually have to wait another generation until it arrives into a building. And we are trying to change that, step by step."

"It usually makes sense to become more intelligent in any human activity," Stulz told me. "As the former Saudi Arabian oil minister Sheikh Yamani once said, the Stone Age didn't end because there were no more stones. It ended because people became more intelligent."

WHAT WOULD IT take to lead a two-thousand-watt life? When I posed this question to Stulz, he gave me another research paper, which offers case studies of six fictionalized households. The Jeannerets are an imaginary family of four who live in Glattbrugg, a town north of Zurich. They own an energy-efficient house, travel by electric bike or train, and occasionally rent a car—they belong to a car-sharing service—to do their grocery shopping. The Moeris, fictional farmers who live northeast of Bern, generate their own electricity with natural gas produced from cow manure; and Alain, Michel, Angela, and Marlène, fictional students living in Geneva, share all their appliances, use the tram, and like to go hiking in the French Alps during school breaks. "There is no formula for how to achieve a two-thousand-watt society," the paper declares. "Three things are needed: societal decisions . . . technical innovation, and the resolve of every individual to act in an energy-conscious way."

Very broadly speaking, the average Swiss today uses energy as follows: fifteen hundred watts per day for living and office space (this includes heat and hot water), eleven hundred watts for food and consumer items (the energy that it takes to produce and transport goods is referred to as "embodied" or "gray" energy), six hundred watts for electricity, five hundred watts for automobile travel, two hundred and fifty watts for air travel, and a hundred and fifty watts for public transportation. Each person's share of Switzerland's public infrastructure, which includes facilities like water- and sewage-treatment plants, comes to nine hundred watts. Reducing these five thousand watts to two thousand would seem to require a significant reduction in every realm. Assuming that infrastructure-related consumption could be cut to five hundred watts, a person who continued to use fifteen hundred watts for living and office space would have nothing left for food, electricity, and transportation. Similarly, a person who continued to travel and use electricity at current rates would consume two thousand watts without having anywhere to live or work, or anything to eat.

While I was in Switzerland, I kept looking for people who actually led two-thousand-watt lives.

"I'm pretty close, except for this stupid air travel," Gerhard Schmitt, the vice president for planning and logistics at the Zurich campus of the Swiss Federal Institute of Technology, told me. "I go once to Shanghai and it's gone." (A round-trip flight between Zurich and Shanghai is the equivalent of using something like eight hundred watts continuously for a year.)

"Let's skip that question," Stulz said when I put it to him. While he lives in an energy-efficient apartment, he, too, travels a great deal; when I visited, he had just returned from a conference in New Delhi, a round trip that used roughly the equivalent of six hundred watts for the year.

The one person I spoke to who did seem to be leading a two-thousand-watt life, or something very near to it, was an engineer named Robert Uetz. Uetz works in the same building as Stulz, and when we returned from visiting the EAWAG Center he was still in his office, even though it was after six. Stulz encouraged me to go talk to him.

"We don't experience it as a restriction," Uetz told me of his two-thousand-watt lifestyle. "On the contrary. I don't feel that we're giving up anything." Uetz and his wife, a dentist, live with their two children in the city of Winterthur, near Zurich. About ten years ago, they bought a two-thousand-square-foot house in a newly built energy-efficient development. The house is heated with a geothermal heat pump—"It's crazy to heat a house with fossil fuels," Uetz said—and has a solar hot-water system. Uetz added photovoltaic panels to the roof to produce electricity; in the winter the panels produce somewhat less power than the house uses—it's equipped with the most energy-efficient lights and appliances the family could find—and in the summer they produce somewhat more, so that over the course of the year the house's electricity use nets out to zero.

"The most important decision was that we wouldn't have a car," Uetz told me. "That was a conscious decision. We looked for a house where we didn't need a car." Driving a lot—even in what, by today's standards at least, counts as an energy-efficient vehicle—also makes it difficult to live within two thousand watts. A person who drives a Toyota Prius ten thousand miles a year consumes roughly two hundred and twenty-five gallons of gasoline. This is equivalent to consuming around eight thousand kilowatt-hours, or to using nearly a thousand watts on a continuous basis. (For a family of four, the same gasoline consumption would come to almost two hundred and fifty watts per person.)

"It's a matter of what you're used to, but I find taking the train a lot more pleasant than driving," Uetz went on. "On the train I can work and relax. If I took a car, I'd have to worry about parking and traffic, rain, snow, and a certain number of people who can't drive but are on the road anyway." When Uetz and his family go on vacation, they travel by rail. "The only thing I'd say that is sort of a restriction is the flying," he said. "Because obviously, with the train where you can go is limited. We can't go to China, or if we did it would take a week.

"I don't make a religion out of it," he added. "I wouldn't do it if I didn't feel good about it—it's how I like to live."

BY THE 2,000-WATT SOCIETY'S own reckoning, cutting consumption is just half—or, perhaps more accurately, a quarter—of what needs to be done. The project's ultimate goal is a world where people consume no more than two thousand watts apiece and where fifteen hundred of those watts come from carbon-free sources. In such a world, everyone would use energy sparingly, like Robert Uetz, and generate it renewably, like Jørgen Tranberg. In such a world, filled with windmills and net-zero houses, carbon emissions would fall sharply, and the concentration of CO_2 in the atmosphere would slowly level off. But how realistic is such a scenario?

Before I left Switzerland to fly back to New York (a trip equivalent to using roughly two hundred and fifty watts continuously for a year), I went to speak to the president of the research council of the Swiss National Science Foundation, Dieter Imboden. Imboden, who is sixty-four, is a compact man with an oval face and silvery hair. He received his training in theoretical solid-state physics, later became interested in environmental physics, and for several years chaired the Swiss Federal Institute of Technology's Environmental Sciences Department. In the late nineties, he served as the director of the 2,000-Watt Society. He said that as a scientist he could see no technical barriers to creating a two-thousand-watt world.

"We are putting our mental energy into the wrong basket," he told me. "Nothing has to be reinvented—for an engineer it's not even a challenge.

"The problems of the twenty-first century are a different kind of problem," he went on. "And I think our society will be measured according to the solution of this new kind of problem, which cannot be solved with the same recipe as the flight to the moon, or the Manhattan Project. It's a qualitative difference—a paradigm change in the role of science for our society."

He continued, "The difficult thing is what I call 'constructed Switzerland.' You in America could call it 'constructed United States'—the buildings and how they are built, but also where they are built and, even more important, the roads, the railroads, the lines for energy, for wastewater, and so on. It's not economically feasible to replace everything in one instant." But since infrastructure should in any case be replaced at the rate of roughly 2 percent a year, if the project is approached incrementally, it's a different task. Then, Imboden said, "it suddenly is feasible."

As of yet, no one has undertaken a rigorous analysis of the economics of a transition to two thousand watts. Researchers have tended, rather, to focus on the price of stabilizing carbon dioxide levels in the atmosphere at a given concentration—either, say, five hundred and fifty parts per million, which is double preindustrial levels, or four hundred and fifty parts, which, many climate scientists say, is the very highest level advisable. Perhaps the most often cited economic study is *The Stern Review,* commissioned by the British government and named for its lead author, Sir Nicholas Stern, formerly the chief economist for the World Bank. *The Stern Review,* published in October 2006, concluded that greenhouse gas levels could be stabilized below double preindustrial concentrations at a cost to global GDP of around 1 percent a year. (*The Stern Review* considered not just CO_2 but other greenhouse gases, like methane and nitrous oxide, as well.) An analysis released last year by the Swedish utility Vattenfall, with research assistance from the American consulting firm McKinsey & Company, reached a similar conclusion: it determined that many measures to reduce carbon emissions, like improving building insulation, would save money, while others, like installing wind turbines, would carry a price. The Vattenfall report estimates that "if all low-cost opportunities are addressed," CO_2 levels could be stabilized at four hundred and fifty parts per million with an annual expenditure of six-tenths of 1 percent of global GDP.

Though 1 percent of the global economy is clearly a lot of money, in the grand scheme of things it's also clearly manageable. It is about a ninth of what's currently spent on healthcare, a seventh of what's spent on oil, and half of what's spent on defense. (More than 40 percent of all the world's military expenditures are made by the United States.) Perhaps most pertinent, it's a far smaller figure than the cost of inaction. *The Stern Review* projects that if current emissions trends are allowed to continue, the eventual damage from climate change will "be equivalent to losing at least 5% of global GDP each year, now and forever," and that "if a wider range of risks and impacts is taken into account" that figure could "rise to 20% of GDP or more."

Twenty years ago, NASA's chief climate scientist, James Hansen, testified on Capitol Hill about the dangers of global warming. Just a few days ago, Hansen returned to the Hill to testify again. "Now, as then, frank assessment of scientific data yields conclusions that are shocking to the body politic," he said. "Now, as then, I can assert that these conclusions have a certainty exceeding 99 percent. The difference is that now we have used up all slack in the schedule." Hansen went on to warn that there would be no practical way to prevent "disastrous" climate change unless the next president and Congress act quickly to curb emissions. Few parts of the U.S. may be as windy as Samsø, or as well organized as Switzerland, but just about everywhere there are possibilities for generating energy more inventively and using it more intelligently. Realizing these possibilities will require a great deal of effort. We may well decide not to make this effort. Such a choice to put off change, however, will merely drive us toward it.

Originally published in *The New Yorker*,
July 7 and 14, 2008.

Update: Samsø now has a large electric car fleet and recently introduced an electric ferry. Meanwhile, annual global emissions have climbed to forty billion tons, and atmospheric CO_2 has risen to more than four hundred and twenty parts per million.

THE SIEGE OF MIAMI

As Temperatures Rise, So, Too, Will Sea Levels

THE CITY OF Miami Beach floods on such a predictable basis that if, out of curiosity or sheer perversity, a person wants to she can plan a visit to coincide with an inundation. Knowing the tides would be high around the time of the "super blood moon," in late September, I arranged to meet up with Hal Wanless, the chairman of the University of Miami's geological sciences department. Wanless, who is seventy-three, has spent nearly half a century studying how South Florida came into being. From this, he's concluded that much of the region may have less than half a century more to go.

We had breakfast at a greasy spoon not far from Wanless's office, then set off across the MacArthur Causeway. (Out-of-towners often assume that Miami Beach is part of Miami, but it's a separate city, on an island a few miles off the coast.) It was a hot, breathless day, with a brilliant blue sky. Wanless turned onto a side street, and soon we were confronting a pond-size puddle. Water gushed down the road and into an underground garage. We stopped in front of a four-story apartment building, which was surrounded by a groomed lawn. Water seemed to be bubbling out of the turf. Wanless took off his shoes and socks and pulled on a pair of polypropylene booties. As he stepped out of the car, a woman rushed over. She asked if he worked for the city. He said he did not, an answer that seemed to disappoint but not deter her. She gestured at a palm tree that was sticking out of the drowned grass.

"Look at our yard, at the landscaping," she said. "That palm tree was super-expensive." She went on, "It's crazy—this is saltwater."

"Welcome to rising sea levels," Wanless told her.

According to the Intergovernmental Panel on Climate Change, sea levels could rise by more than three feet by the end of this century. The U.S. Army Corps of Engineers projects that they could rise by as much as five feet; the National Oceanic and Atmospheric Administration predicts up to six and a half feet. According to Wanless, all these projections are probably low. In his office, Wanless keeps a jar of meltwater he collected from the Greenland ice sheet. He likes to point out that there is plenty more where that came from.

"Many geologists, we're looking at the possibility of a ten- to thirty-foot range by the end of the century," he told me.

We got back into the car. Driving with one hand, Wanless shot pictures out the window with the other. "Look at that," he said. "Oh, my gosh!" We'd come to a neighborhood of multi-million-dollar homes where the water was creeping under the security gates and up the driveways. Porsches and Mercedeses sat flooded up to their chassis.

"This is today, you know," Wanless said. "This isn't with two feet of sea-level rise." He wanted to get better photos, and pulled over onto another side street. He handed me the camera so that I could take a picture of him standing in the middle of the submerged road. Wanless stretched out his arms, like a magician who'd just conjured a rabbit. Some workmen came bouncing along in the back of a pickup. Every few feet, they stuck a depth gauge into the water. A truck from the Miami Beach Public Works Department pulled up. The driver asked if we had called city hall. Apparently, one of the residents of the street had mistaken the high tide for a water-main break. As we were chatting with him, an elderly woman leaning on a walker rounded the corner. She looked at the lake the street had become and wailed, "What am I supposed to do?" The men in the pickup truck agreed to take her home. They folded up her walker and hoisted her into the cab.

To cope with its recurrent flooding, Miami Beach has already spent something like a hundred million dollars. It is planning on spending several hundred million more. Such efforts are, in Wanless's view, so much money down the drain. Sooner or later—and probably sooner—the city will have too much water to deal with. Even before that happens, Wanless believes, insurers will stop selling policies on the luxury condos that line Biscayne Bay. Banks will stop writing mortgages.

"If we don't plan for this," he told me, once we were in the car again, driving toward the Fontainebleau Hotel, "these are the new Okies." I tried to imagine Ma and Pa Joad heading north, their golf bags and espresso machine strapped to the Range Rover.

THE AMOUNT OF water on the planet is fixed (and has been for billions of years). Its distribution, however, is subject to all sorts of re-arrangements. In the coldest part of the last ice age, about twenty thousand years ago, so much water was tied up in ice sheets that sea levels were almost four hundred feet lower than they are today. At that point, Miami Beach, instead of being an island, was fifteen miles from the Atlantic Coast. Sarasota was a hundred miles inland from the Gulf of Mexico, and the outline of the Sunshine State looked less like a skinny finger than like a plump heel.

As the ice age ended and the planet warmed, the world's coastlines assumed their present configuration. There's a good deal of evidence—much of it now submerged—that this process did not take place slowly and steadily but, rather, in fits and starts. Beginning around 12,500 B.C., during an event known as meltwater pulse 1A, sea levels rose by roughly fifty feet in three or four centuries, a rate of more than a foot per decade. Meltwater pulse 1A, along with pulses 1B, 1C, and 1D, was, most probably, the result of ice-sheet collapse. One after another, the enormous glaciers disintegrated and dumped their contents into the oceans. It's been speculated—though the evidence is sketchy—that a sudden flooding of the Black Sea toward the end of meltwater pulse 1C, around seventy-five hundred years ago, inspired the deluge story in Genesis.

As temperatures climb again, so, too, will sea levels. One reason for this is that water, as it heats up, expands. The process of thermal expansion follows well-known physical laws, and its impact is relatively easy to calculate. It is more difficult to predict how the earth's remaining ice sheets will behave, and this difficulty accounts for the wide range in projections.

Low-end forecasts, like the IPCC's, assume that the contribution from the ice sheets will remain relatively stable through the end of the century. High-end projections, like NOAA's, assume that ice-melt will accelerate as the earth warms (as, under any remotely plausible scenario, the planet will continue to do at least through the end of this century, and probably beyond). Recent observations, meanwhile, tend to support the most worrisome scenarios.

The latest data from the Arctic, gathered by a pair of exquisitely sensitive satellites, show that in the past decade Greenland has been losing more ice each year. In August, NASA announced that, to supplement the satellites, it was launching a new monitoring program called—provocatively—Oceans Melting Greenland, or OMG. In November, researchers reported that, owing to the loss of an ice shelf off northeastern Greenland, a new "floodgate" on the ice sheet had opened. All told, Greenland's ice holds enough water to raise global sea levels by twenty feet.

At the opposite end of the earth, two groups of researchers—one from NASA's Jet Propulsion Lab and the other from the University of Washington—concluded last year that a segment of the West Antarctic ice sheet has gone into "irreversible decline." The segment, known as the Amundsen Sea sector, contains enough water to raise global sea levels by four feet, and its melting could destabilize other parts of the ice sheet, which hold enough ice to add ten more feet. While the "decline" could take centuries, it's also possible that it could be accomplished a lot sooner. NASA is already planning for the day when parts of the Kennedy Space Center, on Florida's Cape Canaveral, will be underwater.

THE DAY I TOURED Miami Beach with Hal Wanless, I also attended a panel discussion at the city's convention center titled "Eyes on the Rise." The discussion was hosted by the French government, as part of the lead-up to the climate convention in Paris, at that point two months away. Among the members of the panel was a French scientist named Eric Rignot, a professor at the University of California, Irvine. Rignot is one of the researchers on OMG, and in a conference call with reporters during the summer he said he was "in awe" of how fast the Greenland ice sheet was changing. I ran into him just as he was about to go onstage.

"I'm going to scare people out of this room," he told me. His fellow panelists were a French geophysicist, a climate scientist from the University of Miami, and Miami Beach's mayor, Philip Levine. Levine was elected in 2013, after airing a commercial that tapped into voters' frustration with the continual flooding. It showed him preparing to paddle home from work in a kayak.

"Some people get swept into office," Levine joked when it was his turn at the mike. "I always say I got floated in." He described the steps his administration was taking to combat the effects of rising seas. These include installing enormous underground pumps that will suck water off the streets and dump it into Biscayne Bay. Six pumps have been completed, and fifty-four more are planned. "We had to raise people's stormwater fees to be able to pay for the first hundred-million-dollar tranche," Levine said. "So picture this: you get elected to office and the first thing you tell people is 'By the way, I'm going to raise your rates.'"

He went on, "When you are doing this, there's no textbooks, there's no 'How to Protect Your City from Sea Level Rise,' go to chapter 4." So the city would have to write its own. "We have a team that's going to get it done, that's going to protect this city," the mayor said. "We can't let investor confidence, resident confidence, confidence in our economy start to fall away."

John Morales, the chief meteorologist at NBC's South Florida affiliate, was moderating the discussion. He challenged the mayor, offering a version of the argument I'd heard from Wanless—that today's pumps will be submerged by the seas of tomorrow.

"Down the road, this is just a Band-Aid," Morales said.

"I believe in human innovation," Levine responded. "If, thirty or forty years ago, I'd told you that you were going to be able to communicate with your friends around the world by looking at your watch or with an iPad or an iPhone, you would think I was out of my mind." Thirty or forty years from now, he said, "We're going to have innovative solutions to fight back against sea-level rise that we cannot even imagine today."

MANY OF THE world's largest cities sit along a coast, and all of them are, to one degree or another, threatened by rising seas. Entire countries are endangered—the Maldives, for instance, and the Marshall Islands. Globally, it's estimated that a hundred million people live within three feet of mean high tide and another hundred million or so live within six feet of it. Hundreds of millions more live in areas likely to be affected by increasingly destructive storm surges.

Against this backdrop, South Florida still stands out. The region has been called "ground zero when it comes to sea-level rise." It has also been described as "the poster child for the impacts of climate change," the "epicenter for studying the effects of sea-level rise," a "disaster scenario," and "the New Atlantis." Of all the world's cities, Miami ranks second in terms of assets vulnerable to rising seas—No. 1 is Guangzhou—and in terms of population it ranks fourth, after Guangzhou, Mumbai, and Shanghai. A recent report on storm surges in the United States listed four Florida cities among the eight most at risk. (On that list, Tampa came in at No. 1.) For the past several years, the daily high-water mark in the Miami area has been racing up at the rate of almost an inch a year, nearly ten times the rate of average global sea-level rise. It's unclear exactly why this is happening, but it's been speculated that it has to do with changes in ocean currents that are causing water to pile up along the coast. Talking about climate change in the Everglades this past Earth Day, President Obama said, "Nowhere is it going to have a bigger impact than here in South Florida."

The region's troubles start with its topography. Driving across South Florida is like driving across central Kansas, except that South Florida is greener and a whole lot lower. In Miami-Dade County, the average elevation is just six feet above sea level. The county's highest point, aside from man-made structures, is only about twenty-five feet, and no

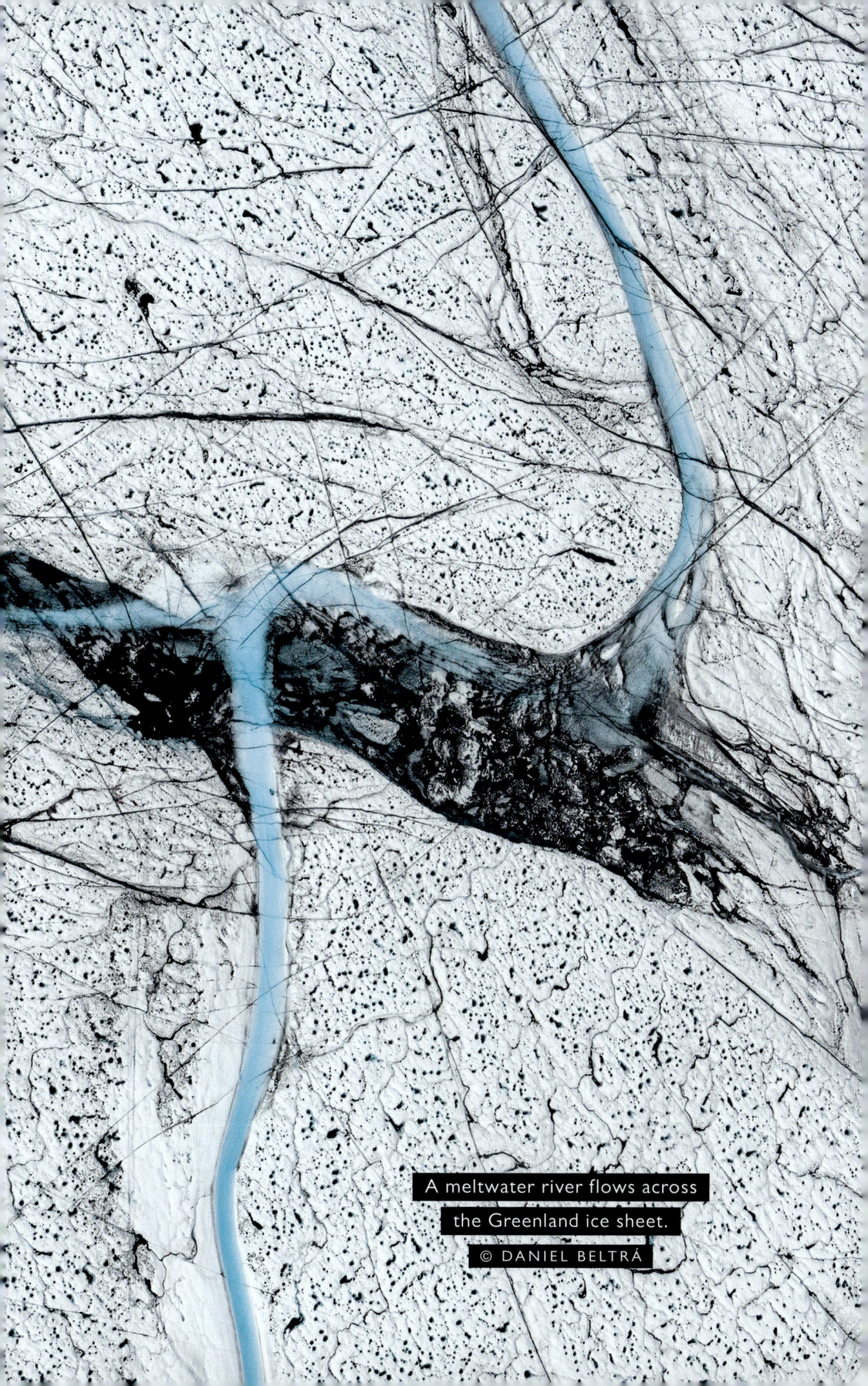

A meltwater river flows across the Greenland ice sheet.

Sirocco, the rock-star kakapo,
New Zealand.

Ursia furtiva caterpillar (middle)
and an unidentified new genus discovered
by Kelsey Wogan (above), Texas.

A pod of sperm whales, Dominica.
TONY WU/NATURE PICTURE LIBRARY

The "bathtub ring" is clearly visible around Lake Powell, Utah.

Sam Wasser sorts through seized tusks, Vietnam.

HOANG DINH NAM/GETTY IMAGES

Sunny-day flooding in Miami Beach, Florida.

LYNNE SLADKY/AP

Christiana Figueres speaking at COP21, Paris, France.

NURPHOTO/GETTY IMAGES

James Hansen getting arrested, Washington, D.C.

JAY MALLIN/ZUMA PRESS, INC./ALAMY

Horses grazing in the Oostvaardersplassen, the Netherlands.
GERT HARDEMAN

Some of RIPE's test plots, Champaign-Urbana, Illinois.
CLAIRE BENJAMIN/UNIVERSITY OF ILLINOIS

one seems entirely sure where it is. (The humorist Dave Barry once set out to climb Miami-Dade's tallest mountain, and ended up atop a local garbage dump nicknamed Mt. Trashmore.) Broward County, which includes Fort Lauderdale, is equally flat and low, and Monroe County, which includes the Florida Keys, is even more so.

But South Florida's problems also run deeper. The whole region—indeed, most of the state—consists of limestone that was laid down over the millions of years Florida sat at the bottom of a shallow sea. The limestone is filled with holes, and the holes are, for the most part, filled with water. (Near the surface, this is generally freshwater, which has a lower density than saltwater.)

Until the eighteen-eighties, when the first channels were cut through the region by steam-powered dredges, South Florida was one continuous wetland—the Everglades. Early efforts to drain the area were only half successful; northerners lured by turn-of-the-century real estate scams found the supposedly rich farmland they'd purchased was more suitable for swimming.

"I have bought land by the acre, and I have bought land by the foot; but, by God, I have never before bought land by the gallon," one arrival from Iowa complained.

Even today, with the Everglades reduced to half its former size, water in the region is constantly being shunted around. The South Florida Water Management District, a state agency, claims that it operates the "world's largest water control system," which includes twenty-three hundred miles of canals, sixty-one pump stations, and more than two thousand "water control structures." Floridians south of Orlando depend on this system to prevent their lawns from drowning and their front steps from becoming docks. (Basement flooding isn't an issue in South Florida, because no one has a basement—the water table is too high.)

When the system was designed—redesigned, really—in the nineteen-fifties, the water level in the canals could be maintained at least a foot and a half higher than the level of high tide. Thanks to this difference in elevation, water flowed off the land toward the sea. At the same time, there was enough freshwater pushing out to prevent saltwater from pressing in. Owing in part to sea-level rise, the gap has since been cut by about eight inches, and the region faces the discomfiting prospect that, during storms, it will be inundated not just along the coasts but also inland, by rainwater that has nowhere to go. Researchers at Florida Atlantic University have found that with just six more inches of sea-level rise the district will lose almost half its flood-control capacity. Meanwhile, what's known as the saltwater front is advancing. One city—Hallandale Beach, just north of Miami—has already had to close most of its drinking wells, because the water is too salty. Many other cities are worried that they will have to do the same.

Jayantha Obeysekera is the Water Management District's chief modeler, which means it's his job to foresee South Florida's future. One morning, I caught up with him at a flood-control structure known as S13, which sits on a canal known as C11, west of Fort Lauderdale.

"We have a triple whammy," he said. "One whammy is sea-level rise. Another whammy is the water table comes up higher too. And in this area the higher the water table, the less space you have to absorb stormwater. The third whammy is if the rainfall extremes change and become more extreme. There are other whammies probably that I haven't mentioned. Someone said the other day, 'The water comes from six sides in Florida.'"

A MONTH AFTER the super blood moon, South Florida experienced another series of very high tides—"king tides," as Miamians call them. This time, I went out to see the effects with Nicole Hernandez Hammer, an environmental studies researcher who works for the Union of Concerned Scientists. Hammer had looked over elevation maps and decided that Shorecrest, about five miles north of downtown Miami, was a neighborhood where we were likely to find flooding. It was another hot, blue morning, and as we drove along, in Hammer's Honda, at first it seemed that she'd miscalculated. Then, all of a sudden, we arrived at a major intersection that was submerged. We parked and made our way onto a side street, also submerged. We were standing in front of a low-slung apartment building, debating what to do next, when one of the residents came by.

"I've been trying to figure out: Where is the water coming from?" he said. "It'll be drying up and then it'll be just like this again." He had complained to the building's superintendent. "I told him, 'Something needs to be done about this water, man.' He says he'll try to do something." A cable repair truck trailing a large wake rolled by and then stalled out.

The water on the street was so deep that it was, indeed, hard to tell where it was coming from. Hammer explained that it was emerging from the storm drains. Instead of funneling rainwater into the bay, as they were designed to do, the drains were directing water from the bay onto the streets. "The infrastructure we have is built for a world that doesn't exist anymore," she said.

Neither of us was wearing boots, a fact that, as we picked our way along, we agreed we regretted. I couldn't help recalling stories I'd heard about Miami's antiquated sewer system, which leaks so much raw waste that it's the subject of frequent lawsuits. (To settle a suit brought by the federal government, the county recently agreed to spend $1.6 billion to upgrade the system, though many question whether the planned repairs adequately account for sea-level rise.) Across the soaked intersection, in front of a single-family home, a middle-aged man was unloading groceries from his car. He, too, told us he didn't know where the water was coming from.

"I heard on the news it's because the moon turned red," he said. "I don't have that much detail about it." During the past month, he added, "it's happened very often." (In an ominous development, Miami this past fall experienced several very high tides at times of the month when, astronomically speaking, it shouldn't have.)

"Honestly, sometimes, when I'm talking to people, I think, Oh, I wish I had taken more psychology courses," Hammer told me. A lot of her job involves visiting low-lying neighborhoods like Shorecrest, helping people understand what they're seeing. She shows them elevation maps and climate-change projections, and explains that the situation is only going to get worse. Often, Hammer said, she feels like a doctor: "You hear that they're trying to teach these skills in medical schools, to encourage them to have a better bedside manner. I think I might try to get that kind of training, because it's really hard to break bad news."

It was garbage collection day, and in front of one house county-issued trash bins bobbed in a stretch of water streaked with oil. Two young women were surveying the scene from the driveway, as if from a pier.

"It's horrible," one of them said to us. "Sometimes the water actually smells." They were sisters, originally from Colombia. They wanted to sell the house, but, as the other sister observed, "No one's going to want to buy it like this."

"I have called the City of Miami," the first sister said. "And they said it's just the moon. But I don't think it's the moon anymore."

After a couple of minutes, their mother came out. Hammer, who was born in Guatemala, began chatting with her in Spanish. "Oh," I heard the mother exclaim. "*Dios mío! El cambio climático!*"

MARCO RUBIO, Florida's junior senator, who has been running third in Republican primary polls, grew up not far from Shorecrest, in West Miami, which sounds like it's a neighborhood but is actually its own city. For several years, he served in Florida's House of Representatives, and his district included Miami's flood-vulnerable airport. Appearing this past spring on *Face the Nation,* Rubio was asked to explain a statement he had made about climate change. He offered the following: "What I said is, humans are not responsible for climate change in the way some of these people out there are trying to make us believe, for the following reason: I believe that climate is changing because there's never been a moment where the climate is not changing."

Around the same time, it was revealed that aides to Florida's governor, Rick Scott, also a Republican, had instructed state workers not to discuss climate change, or even to use the term. The Scott administration, according to the Florida Center for Investigative Reporting, also tried to ban talk of sea-level rise; state employees were supposed to speak, instead, of "nuisance flooding." Scott denied having imposed any such Orwellian restrictions, but I met several people who told me they'd bumped up against them. One was Hammer, who, a few years ago, worked on a report to the state about threats to Florida's transportation system. She said that she was instructed to remove all climate change references from it. "In some places, it was impossible," she recalled. "Like when we talked about the Intergovernmental Panel on Climate Change, which has 'climate change' in the title."

Scientists who study climate change (and the reporters who cover them) often speculate about when the partisan debate on the issue will end. If Florida is a guide, the answer seems to be never. During September's series of king tides, former vice president Al Gore spent a morning sloshing through the flooded streets of Miami Beach with Mayor Levine, a Democrat. I met up with Gore the following day, and he told me that the boots he'd worn had turned out to be too low; the water had poured in over the top.

"When the governor of the state is a full-out climate denier, the irony is just excruciatingly painful," Gore observed. He said that he thought Florida ought to "join with the Maldives and some of the small island states that are urging the world to adopt stronger restrictions on global-warming pollution."

Instead, the state is doing the opposite. In October, Florida filed suit against the Environmental Protection Agency, seeking to block new rules aimed at limiting warming by reducing power plant emissions. (Two dozen states are participating in the lawsuit.)

"The level of disconnect from reality is pretty profound," Jeff Goodell, a journalist who's working on a book on the impacts of sea-level rise, told me. "We're sort of used to that in the climate world. But in Florida there are real consequences. The water is rising right now."

Meanwhile, people continue to flock to South Florida. Miami's metropolitan area, which includes Fort Lauderdale, has been one of the fastest growing in the country; from 2013 to 2014, in absolute terms it added more residents than San Francisco and, proportionally speaking, it outdid Los Angeles and New York. Currently, in downtown Miami there are more than twenty-five thousand new condominium units either proposed or under construction. Much of the boom is being financed by "flight capital" from countries like Argentina and Venezuela; something like half of recent home sales in Miami were paid for in cash.

And just about everyone who can afford to buys near the water. Not long ago, Kenneth Griffin, a hedge fund billionaire, bought a penthouse in Miami Beach for sixty million dollars, the highest amount ever paid for a single-family residence in Miami-Dade County (and ten million dollars more than the original asking price). The penthouse, in a new building called Faena House, offers eight bedrooms and a seventy-foot rooftop pool. When I read about the sale, I plugged the building's address into a handy program called the Sea Level Rise Toolbox, created by students and professors at Florida International University. According to the program, with a little more than one foot of rise the roads around the building will frequently flood. With two feet, most of the streets will be underwater, and with three it seems that, if Faena House is still habitable, it will be accessible only by boat.

I ASKED EVERYONE I met in South Florida who seemed at all concerned about sea-level rise the same question: What could be done? More than a quarter of the Netherlands is below sea level, and those areas are home to millions of people, so low-elevation living is certainly possible. But the geology of South Florida is peculiarly intractable. Building a dike on porous limestone is like putting a fence on top of a tunnel: it alters the route of travel, but not necessarily the amount.

"You can't build levees on the coast and stop the water" is the way Jayantha Obeysekera put it. "The water would just come underground."

Some people told me that they thought the only realistic response for South Florida was retreat.

"I live opposite a park," Philip Stoddard, the mayor of South Miami—also a city in its own right—told me. "And there's a low area in it that fills up when it rains. I was out there this morning walking my dog, and I saw fish in it. Where the heck did the fish come from? They came from underground. We have fish that travel underground!

"What that means is, there's no keeping the water out," he went on. "So ultimately this area has to depopulate. What I want to work toward is a slow and graceful depopulation, rather than a sudden and catastrophic one."

More often, I heard echoes of Mayor Levine's Apple Watch line. Who knows what amazing breakthroughs the future will bring?

"I think people are underestimating the incredible innovative imagination in the world of adaptive design," Harvey Ruvin, the clerk of the courts of Miami-Dade County and the chairman of the county's Sea Level Rise Task Force, said when I went to visit him in his office. A quote from Buckminster Fuller hung on the wall: "We are all passengers on Spaceship Earth." Ruvin became friendly with Fuller in the nineteen-sixties, after reading about a plan Fuller had drawn up for a floating city in Tokyo Bay.

"I would agree that things can't continue exactly the way they are today," Ruvin told me. "But what we will evolve to may be better."

"I KEEP TELLING PEOPLE, 'This is my patient,'" Bruce Mowry, Miami Beach's city engineer, was saying. "I can't lose my patient. If I don't do anything, Miami Beach may not be here." It was yet another day of bright-blue skies and "nuisance flooding," and I was walking with Mowry through one of Miami Beach's lowest neighborhoods, Sunset Harbour.

If Miami Beach is on a gurney, then Mowry might be said to be thumping its chest. It's his job to keep the city viable, and since no one has yet come up with a smart-watch-like breakthrough, he's been forced to rely on more primitive means, like pumps and asphalt. We rounded a corner and came to a set of stairs, which led down to some restaurants and shops. Until recently, Mowry explained, the shops and the street had been at the same level. But the street had recently been raised. It was now almost a yard higher than the sidewalk.

"I call this my five-step program," he said. "What are the five steps?" He counted off the stairs as we descended: "One, two, three, four, five." Some restaurants had set up tables at the bottom, next to what used to be a curb but now, with the elevation of the road, is a three-foot wall. Cars whizzed by at the diners' eye level. I found the arrangement disconcerting, as if I'd suddenly shrunk. Mowry told me that some of the business owners, who had been unhappy when the street flooded, now were unhappy because they had no direct access to the road: "It's, like, can you win?"

Several nearby streets had also been raised, by about a foot. The elevated roadbeds were higher than the driveways, which now all sloped down. The parking lot of a car-rental agency sat in a kind of hollow.

I asked about the limestone problem. "That is the one that scares us more than anything," Mowry said. "New Orleans, the Netherlands—everybody understands putting in barriers, perimeter levees, pumps. Very few people understand: What do you do when the water's coming up through the ground?

"What I'd really like to do is pick the whole city up, spray on a membrane, and drop it back down," he went on. I thought of Italo Calvino's *Invisible Cities,* where such fantastical engineering schemes are the norm.

Mowry said he was intrigued by the possibility of finding some kind of resin that could be injected into the limestone. The resin would fill the holes, then set to form a seal. Or, he suggested, perhaps one day the city would require that builders, before constructing a house, lay a waterproof shield underneath it, the way a camper spreads a tarp under a tent. Or maybe some sort of clay could be pumped into the ground that would ooze out and fill the interstices.

"Will it hold?" Mowry said of the clay. "I doubt it. But these are things we're exploring." It was hard to tell how seriously he took any of these ideas; even if one of them turned out to be workable, the effort required to, in effect, caulk the entire island seemed staggering. At one point, Mowry declared, "If we can put a man on the moon, then we can figure out a way to keep Miami Beach dry." At another, he mused about the city's reverting to "what it came from," which was largely mangrove swamp: "I'm sure if we had poets, they'd be writing about the swallowing of Miami Beach by the sea."

We headed back toward Mowry's office around the time of maximum high tide. The elevated streets were still dry, but on the way to city hall we came to an unreconstructed stretch of road that was flooding. Evidently, this situation had been anticipated, because two mobile pumps, the size and shape of ice cream trucks, were parked near the quickly expanding pool. Neither was operating. After making a couple of phone calls, Mowry decided that he would try to switch them on himself. As he fiddled with the controls, I realized that we were standing not far from the drowned palm tree I'd seen on my first day in Miami Beach, and that it was once again underwater.

ABOUT A DOZEN miles due west of Miami, the land gives out, and what's left of the Everglades begins. The best way to get around in this part of Florida is by airboat, and on a gray morning I set out in one with a hydrologist named Christopher McVoy. We rented the boat from a concession run by members of the Miccosukee Tribe, which, before the Europeans arrived, occupied large swaths of Georgia and Tennessee. The colonists hounded the Miccosukee ever farther south until, eventually, they ended up with a few hundred mostly flooded square miles between Miami and Naples. On a fence in front of the dock, a sign read, "Beware: Wild alligators are dangerous. Do not feed or tease." Our guide, Betty Osceola, handed out headsets to block the noise of the rotors, and we zipped off.

The Everglades is often referred to as a "river of grass," but it might just as accurately be described as a prairie of water. Where the airboats had made a track, the water was open, but mostly it was patchy—interrupted by clumps of sawgrass and an occasional tree island. We hadn't been out very long when it started to pour. As the boat sped into the rain, it felt as if we were driving through a sandstorm.

The same features that now make South Florida so vulnerable—its flatness, its high water table, its heavy rains—are the features that brought the Everglades into being. Before the drainage canals were dug, water flowed from Lake Okeechobee, about seventy miles north of Miami, to Florida Bay, about forty miles to the south of the city, in one wide, slow-moving sheet. Now much of the water is diverted, and the water that does make it to the wetlands gets impounded, so the once-continuous "sheet flow" is no more. There's a comprehensive Everglades restoration plan, which goes by the acronym CERP, but this has gotten hung up on one political snag after another, and climate change adds yet one more obstacle. The Everglades is a freshwater ecosystem; already, at the southern margin of Everglades National Park, the water is becoming salty. The sawgrass is in retreat, and mangroves are moving in. In coming decades, there's likely to be more and more demand for the freshwater that remains. As McVoy put it, "You've got a big chunk of agriculture, a big chunk of people, and a big chunk of nature reserve all competing for the same resources."

The best that can be hoped for with the restoration project is that it will prolong the life of the wetland and, with that, of Miami's drinking-water system. But you can't get around geophysics. Send the ice sheets into "irreversible decline," as it seems increasingly likely we have done, and there's no going back. Eventually, the Everglades, along with Shorecrest and Miami Beach and much of the rest of South Florida, will be inundated. And, if Hal Wanless is right, "eventually" isn't very far off.

To me, the gunmetal expanse of water and grass appeared utterly without markers, but Osceola, who could read the subtlest of ridges, knew exactly where we were at every moment. We stopped to have sandwiches on an island with enough dry land for a tiny farm, and stopped again at a research site that McVoy had set up in the muck. There was a box of electrical equipment on stilts, and a solar panel to provide power. McVoy dropped out of the boat to collect some samples in empty water-cooler bottles. The rain let up and then started again.

Originally published in *The New Yorker,*
December 21 and 28, 2015.

PART THREE

BIG IDEAS

TESTING THE WATERS

Should the Natural World Have Rights?

L AKE MARY JANE is shallow—twelve feet deep at most—but she's well connected. She makes her home in central Florida, in an area that was once given over to wetlands. To the north, she is linked to a marsh, and to the west a canal ties her to Lake Hart. To the south, through more canals, Mary Jane feeds into a chain of lakes that run into Lake Kissimmee, which feeds into Lake Okeechobee. Were Lake Okeechobee not encircled by dikes, the water that flows through Mary Jane would keep pouring south until it glided across the Everglades and out to sea.

Mary Jane has an irregular shape that, on a map, looks a bit like a woman's head in profile. Where the back of the woman's head would be, there's a park fitted out with a playground and picnic tables. Where the face would be, there are scattered houses, with long docks that teeter over the water. People who live along Mary Jane like to go boating and swimming and watch the wildlife. Toward the park side of the lake sits an islet, known as Bird Island, that's favored by nesting egrets and wood storks.

Like most of the rest of central Florida, Mary Jane is under pressure from development. Orange County, which encompasses the lake, the city of Orlando, and much of Disney World, is one of the fastest-growing counties in Florida, and Florida is one of the fastest-growing states in the nation. A development planned for a site just north of

Mary Jane would convert nineteen hundred acres of wetlands, pine flatlands, and cypress forest into homes, lawns, and office buildings.

In an effort to protect herself, Mary Jane is suing. The lake has filed a case in Florida state court, together with Lake Hart, the Crosby Island Marsh, and two boggy streams. According to legal papers submitted in February, the development would "adversely impact the lakes and marsh who are parties to this action," causing injuries that are "concrete, distinct, and palpable."

A number of animals have preceded Mary Jane to court, including Happy, an elephant who lives at the Bronx Zoo, and Justice, an Appaloosa cross whose owner, in Oregon, neglected him. There have also been several cases brought by entire species: for instance, the palila, a critically endangered bird, successfully sued Hawaii's Department of Land and Natural Resources for allowing feral goats to graze on its last remaining bit of habitat. (The palila "wings its way into federal court in its own right," Diarmuid O'Scannlain, a judge on the U.S. Court of Appeals for the Ninth Circuit, wrote in a decision that granted the species relief.)

Still, Mary Jane's case is a first. Never before has an inanimate slice of nature tried to defend its rights in an American courtroom. Depending on your perspective, the lake's case is either borderline delusional or way overdue.

"It is long past time to recognize that we are dependent on nature, and the continued destruction of nature needs to stop," Mari Margil, the executive director of the Center for Democratic and Environmental Rights, said in a statement celebrating the lawsuit.

"Your local lake or river could sue you?" the Florida Chamber of Commerce said. "Not on our watch."

THE NOTION THAT "natural objects" like woods and streams should have rights was first put forward half a century ago, by Christopher Stone, a law professor at the University of Southern California. Stone, who died last year, was a son of the crusading journalist I. F. Stone, and as a kid, in the nineteen-fifties, he sometimes helped put out his father's newspaper, *I. F. Stone's Weekly.* In the fall of 1971, the younger Stone was assigned to teach USC's introductory course on property law, and in one class he delivered a lecture on how ownership rights had evolved over time. Near the end of the hour, sensing that his students' minds were wandering, he decided to shake things up. What would happen, he asked, if the law were to further evolve to grant rights to, say, trees or even rocks? "This little thought experiment," he later recalled, created an "uproar."

Until that moment, Stone hadn't considered this question. But, having tossed it out, he found himself intrigued. He set about writing a law review article. In the article, "Should Trees Have Standing? Toward Legal Rights for Natural Objects," Stone noted that rights are always socially constructed. In America in the eighteenth and nineteenth centuries, many groups—Blacks, Native Americans, women, children—were denied rights; then, as society, or what counted as society, changed, rights were slowly and painfully (and often incompletely) extended to them.

"Each time there is a movement to confer rights onto some new 'entity,' the proposal is bound to sound odd or frightening or laughable," Stone wrote. "This is partly because until the rightless thing receives its rights we cannot see it as anything but a thing for the use of 'us'—those who are holding rights at the time." He went on, "I am quite seriously proposing that we give legal rights to forests, oceans, rivers and other so-called 'natural objects' in the environment—indeed to the natural environment as a whole."

This extension of rights, Stone argued, was needed to address an otherwise insuperable problem. So long as "natural objects" were valued only in terms of their worth to humans—"for the use of 'us'"—they could, quite legally, be destroyed. Stone cited the example of someone polluting a stream. People living downstream could take the polluter to court and perhaps win damages. But the waterway and the species dependent on it would never recoup their losses. In the conflict between the polluter and the downstream residents, he wrote, "the stream itself is lost sight of."

As it happened, in the autumn of 1971, while Stone was at work on his article, a major environmental case was wending its way through the courts. A couple of years earlier, Disney had decided to build a giant ski resort in a wilderness area south of Yosemite known as Mineral King. (The resort was to be, in Disney's words, an "American Alpine Wonderland," with a five-story hotel, twenty-two lifts, and ten restaurants, including one at eleven thousand feet.) To construct the resort, and to bring in visitors, the company needed an access road through Sequoia National Park. When the Interior Department approved the highway, the Sierra Club sued, arguing that it would cause "irreparable harm to the public interest." A federal judge in San Francisco ruled in the group's favor and issued a preliminary injunction blocking work on the resort. On an appeal from the Interior Department, the ruling was reversed. The Sierra Club, the appellate court said, lacked standing to sue, since it wouldn't be directly affected by the project. This time, the Sierra Club appealed.

When Stone learned that the case, *Sierra Club v. Morton,* was headed to the U.S. Supreme Court, he decided, with the help of the editors of the *Southern California Law Review,* to rush his article into print. A friend of his, who was a law clerk for Supreme Court Justice William O. Douglas, seems to have relayed an early draft to Douglas, an ardent environmentalist. (Whether this back-channel communication was kosher is debatable.)

In April 1972, the Supreme Court upheld the appellate court's decision against the Sierra Club, by a vote of four to three. (Two seats on the court were vacant.) Douglas, drawing heavily on Stone's article, penned a dissenting opinion. "A ship has a legal personality, a fiction found useful for maritime purposes," he wrote. A corporation, too, "is a 'person' for purposes of the adjudicatory processes. . . . So it should be as respects valleys, alpine meadows, rivers, lakes, estuaries, beaches, ridges, groves of trees, swampland, or even air that feels the destructive pressures of modern technology and modern life."

Douglas's opinion has been described as "one of the most famous and passionate dissents in the Supreme Court's history," and it turned what probably otherwise would have been a little-noticed law review article into a media event. "Should Trees Have Standing?" was reprinted in the *Congressional Record* and published in book form. *The Berkeley Monthly* declared it a sign of better times to come. There was something "amiably zany," as Stone would later put it, about a law professor who wanted to bestow rights on shrubs.

Even Stone's critics had fun with his idea. "Why wouldn't Mineral King want to host a ski resort, after doing nothing for a billion years?" Mark Sagoff, a philosophy professor, quipped in *The Yale Law Journal*. Writing in the *American Bar Association Journal*, an attorney named John Naff lyricized:

> Great mountain peaks of names prestigious
> Will suddenly become litigious.
> Our brooks will babble in the courts,
> Seeking damages for torts.
> How can I rest beneath a tree
> If it may soon be suing me?

THE BODIES OF water that have filed suit in Orange County have one co-plaintiff who walks on two legs, and that is Chuck O'Neal. O'Neal is sixty-six, with slate-gray hair, a broad face, and a reedy voice. He is the founder of Speak Up Wekiva, an organization named for a river that runs near his home, and until recently he was also the president, the chairman, and the director-at-large of a group called the Florida Rights of Nature Network.

"I often hear the word *radical*," O'Neal told me. "And I'm, like, all right. *Radical* comes from the Latin word *radix,* for 'root,' and that's exactly what this is: change at the root. Does nature have rights? That concept, I agree, is radical."

One morning not long ago, O'Neal picked me up at the hotel where I was staying, north of downtown Orlando. Our plan was to tour all the bodies of water that have filed suit, starting with a stream called Wilde Cypress Branch. The drive took us past strings of shopping centers and clusters of condominiums, then past more shopping centers and the walls of gated communities. Eventually, we arrived at an area that wasn't quite rural but also wasn't quite suburban. O'Neal pulled off the road next to some open land studded with scraggly bushes. Stretching around the bushes, and for as far as I could see, was a five-foot-high barbed-wire fence, which appeared to be new. O'Neal explained that to reach Wilde Cypress we would have to get across the fence. While we were debating how to do this—over or under?—two men in a white pickup truck drove up and parked behind O'Neal's car. The rapidity with which they'd shown up freaked us out, and we decided to head off to see another plaintiff, a stream known as Boggy Branch.

O'Neal, who grew up in Orange County and lives in the town of Apopka, describes himself as a "serial entrepreneur." These days, he runs a business that mostly involves buying houses and flipping them. As we talked, he occasionally received calls on his cell phone from building supply stores. We rode past fields occupied by clusters of black cows. O'Neal speculated that these were "rental cows." In Florida, he explained, land that's being grazed enjoys special tax advantages, which developers often avail themselves of until a parcel can be filled with something more profitable.

After a while, we pulled into a stretch of brand-new, tightly spaced houses, some still being framed. A banner that hung on a construction barrier identified the development as Meridian Parks: the "Perfect Place to Start," it said. Between two groups of homes, the road ended abruptly in a set of reflective warning signs. Just beyond the signs lay Boggy Branch. More swamp than stream, it seemed barely to be moving. Cypress trees festooned with Spanish moss rose out of the black water. A ridge, clearly man-made, separated Boggy Branch from a large retention pond, also clearly man-made.

O'Neal had brought a map of the development he was fighting—a proposed extension of Meridian Parks known, inelegantly, as Meridian Parks Remainder. About a third of the map was stippled with black dots, indicating wetlands. To complete the project, which is supposed to include town houses, apartment buildings, and commercial space, the development company, Beachline South Residential, planned to extend the road across Boggy Branch and then across Wilde Cypress Branch. The roadwork and various other rearrangements of the landscape would entail filling in, or otherwise altering, wetlands covering more than a hundred acres. This was what the bodies of water were suing over. The move, their lawyer argued, would restrict the natural flow from the streams into the lakes, thereby wreaking havoc with the local ecology and threatening the lakes' right to exist.

"This water has been flowing this way for tens of thousands of years," O'Neal said, as we tromped along the ridge, more or less in people's backyards. "Where's that being considered anywhere in this development?"

A FEW YEARS after *Sierra Club v. Morton,* Justice Douglas retired from the Supreme Court. Stone, meanwhile, moved on to other subjects. Like a vernal pool in summer, interest in "Should Trees Have Standing?" started to dry up. Then it bubbled back to life.

In 2005, residents of Tamaqua Borough, in eastern Pennsylvania, were fighting a plan to dump toxic sludge in an open pit in town. One of the members of Tamaqua Borough's town council attended a meeting with representatives of the Community Environmental Legal Defense Fund, which had been set up to help local groups battle such projects. The organization's leader at the time, a lawyer named Thomas Linzey, had recently chanced upon Stone's article. It occurred to him that if trees—or, in the case of Tamaqua, ecosystems more generally—had standing then the town would have another legal tool to use in its campaign. He helped draft a local anti-sludge ordinance that, among many other things, declared it "unlawful for any corporation or its directors . . . to interfere with the existence and flourishing of natural communities." For the purposes of the ordinance, natural communities were to be considered "persons." When the ordinance came up for a vote, in 2006, the town council was split, three to three. Tamaqua's mayor cast the tie-breaking vote, in favor of the "natural communities." He later said, "If I am going to be sued, so be it."

The proposed dump was canceled, so Tamaqua's ordinance, believed to be the first of its kind in the world, was never put to the test. Still, one thing led to another, and a year later Linzey received what he described to me as a "weird phone call." Ecuador had elected a group of delegates to rewrite its constitution, and someone involved in the assembly had somehow heard about the Tamaqua ordinance. Linzey was invited to the Ecuadorian city of Montecristi. He ended up traveling to the country several times to consult with the delegates. "That was pretty cool," he recalled.

When Ecuador's new constitution was adopted, in 2008, it marked another, much more significant world first. The constitution's preamble celebrates Pacha Mama, usually translated from Quechua as "Mother Earth," and a later section enumerates the rights that Pacha Mama enjoys. These include "the right to integral respect for its existence" and "the right to be restored." The constitution also includes a right to *buen vivir,* which translates into English as "good living," but is itself a translation of the Quechua term *sumak kawsay,* which has far-reaching spiritual and political implications.

"Ecuador is a country that takes pluralism very seriously," Hugo Echeverría, an environmental lawyer in Quito, told me. "And the philosophical concept behind the rights of nature fits into the vision of its Indigenous peoples. That's why you find the words Pacha Mama in the constitution." In a recent landmark, or at least land-centric, decision, the country's highest court ruled that mining permits that had been granted in Los Cedros, a protected forest north of Quito, violated the constitution and should be voided. (Most of the permits are held by Ecuador's national mining company, which goes by the acronym ENAMI.)

"Los Cedros is a key case because it applied the constitution in a context where it was difficult to apply," Echeverría said. "Wildlife was chosen over mining, which is a very important activity in Ecuador because it provides income to the state. No court has ever taken that step before."

After his experience in Ecuador, Linzey continued to travel, in the hope of finding more communities—or countries—interested in granting rights to nature. In 2013, he flew to Orlando to speak at a seminar at the Barry University School of Law. Sitting in the audience was Chuck O'Neal. O'Neal, who was active in local environmental causes, was intrigued by what he heard, but he had his doubts. "For Florida, I just didn't think it would work," he told me. Then, in 2018, a toxic algae bloom the size of Connecticut turned the Florida Gulf Coast into a slick of dead fish.

O'Neal put aside his doubts. It was time, he decided, to try something new. In the spring of 2019, he invited Linzey back to Florida, to speak to a group of concerned citizens from around the state. (Soon afterward, Linzey went to work for the Center for Democratic and Environmental Rights, a group that he helped found.) The attendees agreed to go home and try to pass rights-of-nature laws in the regions where they lived.

With the help of some other Orange County residents, O'Neal wrote up a "bill of rights" for the Wekiva River and for the Econlockhatchee, a second river that passes near Orlando. He presented the bill to a commission that had been appointed to revise Orange County's charter. To his surprise, the commission didn't just take up his proposal; it expanded it. In November 2020, when voters went to the polls, they were asked whether all of the county's waterways—be they "fresh, brackish, saline, tidal, surface or underground"—should have the "right to exist, flow, to be protected against pollution and to maintain a healthy ecosystem." Eighty-nine percent voted to approve the charter amendment, which did better than almost anything or anyone else on the ballot in the county, including Joe Biden, who got 61 percent.

The Orlando *Sentinel* said that the amendment had unified voters "in a state with a lousy track record for protecting natural resources." It nominated O'Neal for "Central Floridian of the Year" and dubbed him "our local Lorax."

IN ADDITION TO prompting Mary Jane's lawsuit, the Orange County charter amendment has inspired an art installation, and one day while I was in Florida O'Neal took me to see it. It was being exhibited in a ranch house turned gallery, which was decorated on the outside with wild swirls of tile. We knocked on the front door, and the artist, Brooks Dierdorff, answered.

In what had presumably once been the dining room, a large white slab covered most of the floor. On it rested several documents, including Ecuador's constitution, and several glasses of water. Dierdorff, who teaches photography at the University of Central Florida, explained that the water had been collected from lakes and streams around Orange County. Most of it was clear, but one glassful was the color of strong tea. This turned out to be a sample from Lake Mary Jane, which is naturally high in tannins. O'Neal bent over to peer into the glass. "Wow, that's really dark," he said.

Dierdorff told us that his goal was to visit every lake, river, and stream in the county. Each time he went to a new one, he would add another glass: "My plan is to have things change and evolve over time."

What had once perhaps been the house's den was bathed in violet light and outfitted with speakers. Dierdorff told us that he was also collecting sound: at each waterway he sampled, he stuck a waterproof mike into the depths. He had layered sixteen of these recordings on top of one another, and the resulting track was playing on a loop. "I think of it as kind of a chorus," he said.

I said I couldn't hear anything. Dierdorff shrugged. "There are some little blips once in a while," he said.

As rights holders, natural objects have an obvious deficit: they cannot speak for themselves. Even if granted standing, they have to rely on people to plead their cause. And since it's hard to pull together a jury of, say, jungles, it's people who also have to decide their interests. Animals may in some way be able to convey their desires—or at least allow humans to believe that they can. But, apart from "some little blips," a swamp doesn't communicate much. Who can really claim to know its will?

"It is far from clear that it feels like anything to be an oak tree," Mauricio Guim and Michael Livermore, both law professors, argue in "Where Nature's Rights Go Wrong," an article that appeared recently in the *Virginia Law Review*. "Nor does it feel like anything to be a rainforest ecosystem, even if it is teeming with birds who have some form of subjective experience."

The objection that streams and forests cannot have standing because streams and forests cannot speak was, in Stone's view, easily addressed. "Corporations cannot speak either," he observed. "Nor can states, estates, infants, incompetents, municipalities or universities." And yet these entities were amply represented—some might say overrepresented—in the courts.

"We make decisions on behalf of, and in the purported interests of, others every day," Stone wrote. "These 'others' are often creatures whose wants are far less verifiable, and even far more metaphysical in conception, than the wants of rivers, trees, and land." He envisaged a system of guardianships by which "a friend of a natural object," perceiving it to be endangered, could apply to a court to represent it. The guardian could try to prevent, or demand redress for, injuries that had no quantifiable human cost, such as "the loss from the face of the earth of species of commercially valueless birds" or "the disappearance of a wilderness area."

Start taking Stone seriously and it's hard to stop. From a certain point of view, granting nature a say isn't radical or new at all. For most of history, people saw themselves as dependent on their surroundings, and "rivers, trees, and land" enjoyed the last word. Only in the past few hundred years has it become possible—and come to seem normal—for people to mow down forests, fill in wetlands, and blast away mountains because it suits them. This way of operating has resulted in unprecedented, if unequally distributed, human prosperity. It has also brought melting ice sheets, marine dead zones, soaring extinction rates, and the prospect of global ecological collapse. As António Guterres, the secretary-general of the United Nations, put it last week, when the latest international climate report was released, we are "firmly on track towards an unlivable world."

THERE'S NO WAY to get back to Eden, but it's easy to get to Eden Bar, which advertises itself as "Central Florida's most unique outdoor restaurant." I arranged to meet up there one evening with Steven

Meyers, the attorney working on Mary Jane's lawsuit. He was nursing a glass of red wine when I arrived. He had just filed an eighty-page brief on behalf of the lakes and streams, a copy of which lay on the table.

"I'm a personal-injury lawyer," he told me as soon as I sat down. Mostly he filed workers' comp cases. He had got involved with O'Neal, he explained, because of a dead bear.

In 2015, Rick Scott, then Florida's governor, had reinstituted bear hunting in the state. One night, while Meyers was working late, he came across a video of a black bear being shot in Canada. The gruesomeness of the images—the bear did not immediately die but kept getting up and falling down again—shook him. He had read that O'Neal, whom he'd never met, had filed a lawsuit to try to prevent bear hunting in Florida. Suddenly, Meyers felt moved to get in touch with him.

"That night, I emailed Chuck," Meyers recalled, "and I said, 'I'm not an environmental lawyer or an animal rights lawyer, but here's a donation, and if I can help I'd love to.' I thought I probably would never hear anything. Twenty minutes later, he sends me all the pleadings. He's like, 'Welcome to the team.'"

Meyers ended up working, pro bono, on the bear-hunting case. He couldn't get a judge to stop the hunt, though O'Neal did manage to persuade state wildlife officials to put an end to the shooting after three hundred and four bears were killed in two days. Meyers and O'Neal then worked together on an effort to block a hundred-and-twenty-acre warehouse development planned for a site near the headwaters of the Little Wekiva River. They lost that case and, as a part of the legal settlement, may have to pay three thousand dollars to help defray their opponents' legal costs.

Their latest case, on behalf of Mary Jane et al., also seems likely to fail. As soon as the Orange County Charter Commission decided, in early June 2020, to put the bill of rights for waterways on the ballot, business lobbyists in Tallahassee sprang into action. In a bill that mostly had to do with regulating septic systems, an amendment suddenly appeared prohibiting local governments from granting legal rights to any "part of the natural environment." The state legislature passed the bill in mid-June, and it went into effect in July, meaning that by the time Orange County voted to approve the charter amendment, in November, it had already been preempted.

The developer, Beachline South Residential, is pushing to have Mary Jane's case dismissed, arguing that the rights the lake is invoking do not—and cannot—exist. The state legislature could not have been "clearer in its intention to nullify" the Orange County charter amendment, papers filed by Beachline's legal team note. For their part, the bodies of water, which is to say Meyers and O'Neal, argue that the preemption is itself invalid. In the words of the brief that Meyers brought to Eden Bar, it is "unconstitutional, unlawful, and inapplicable."

"We're realistic," Meyers told me. "We're trying to make new law, and that's always hard. But it's like Michael Jordan said: you miss a hundred percent of the shots you don't take."

I had invited O'Neal to join us, and after a while he showed up at the bar. The conversation turned from Mary Jane's lawsuit to infighting among her allies. The Florida Rights of Nature Network, a group founded in O'Neal's living room, wanted to try to pass another amendment, this one to the Florida state constitution. The hope was to preempt the preemption of laws like Orange County's. O'Neal had one idea about how to word the proposed amendment; other members of the group had a different idea. The argument had become so heated that O'Neal had broken with the group and resigned as its president, chairman, and director-at-large, and also as chair of its political committee.

"One thing about Chuck is he gets along with everyone," Meyers teased, rolling his eyes. It was impossible for me to know whether O'Neal's pique was justified, but it occurred to me, and not for the first time, that a nature dependent on human collegiality was in deep trouble.

The next day, I got up early. It was my last day in Florida, and I wanted to pay Mary Jane another visit before I headed home. When I arrived at the park on her western shore, I had the place pretty much to myself. It was a lovely morning, with a blue sky and a light breeze.

Mary Jane doesn't really have a beach, so I sat down on a patch of more or less dry ground. Sticking out of the soggy grass was a sign that read "Alligators and snakes are common in this area" and, beneath that, "KEEP YOUR DISTANCE."

A wood stork arrived and started poking its beak into the muck at the lake's edge. More storks swooped down and similarly began poking. One of them bent its legs, dipped its white-and-black wings into the water, and then held them out, as if airing a blanket. Another stork did the same, and soon they were all rolling around in the water and stretching their wings. I wasn't sure what, exactly, they were doing, but it looked like fun. I took off my shoes and waded in. As I approached, most of the storks flew away. The water, around my ankles, was the golden brown I had seen in Dierdorff's exhibit. I spent a while listening. I didn't hear any blips from Mary Jane; still, it seemed to me, the lake's wishes were pretty clear, as were the wood storks'. What they really wanted was to be left alone.

Originally published in *The New Yorker*,
April 18, 2022.

A NEW LEAF

Could Tinkering with Photosynthesis
Prevent a Global Food Crisis?

THIS STORY BEGINS about two billion years ago, when the
world, if not young, exactly, was a lot more impressionable. The
planet spun faster, so the sun rose every twenty-one hours. The earliest
continents were forming—Arctica, for instance, which persists as bits
and pieces of Siberia. Most of the globe was given over to oceans, and
the oceans teemed with microbes.

Some of these microbes—the group known as cyanobacteria—had
mastered a peculiarly powerful form of alchemy. They lived off sunlight, which they converted into sugar. As a waste product, they gave
off oxygen. Cyanobacteria were so plentiful, and so good at what they
did, that they changed the world. They altered the oceans' chemistry,
and then the atmosphere's. Formerly in short supply, oxygen became
abundant. Anything that couldn't tolerate it either died off or retreated
to some dark, airless corner.

One day, another organism—a sort of proto-alga—devoured a cyanobacterium. Instead of being destroyed, as you might expect, the bacterium took up residence, like Jonah in the whale. This accommodation,
unlikely as it was, sent life in a new direction. The secret to photosynthesis passed to the alga and all its heirs.

A billion years went by. The planet's rotation slowed. The continents crashed together to form a supercontinent, Rodinia, then drifted
apart again. The alga's heirs diversified.

One side of the family stuck to the water. Another branch set out to colonize dry land. The first explorers stayed small and low to the ground. (These were probably related to liverworts.) Eventually, they were joined by the ancestors of today's ferns and mosses. There was so much empty space—and hence available light—that plants, as one botanist has put it, found terrestrial life "irresistible." They spread out their fronds and began to grow taller. The rise of plants made possible the rise of plant-eating animals. During the Carboniferous period, towering tree ferns and giant club mosses covered the earth, and insects with wingspans of more than two feet flitted through them.

Some two hundred million years later, in the early Cretaceous, plants with flowers appeared on the scene. They were so fabulously successful that they soon took over. (Charles Darwin was deeply troubled by the sudden appearance of flowering plants in the fossil record, describing it as an "abominable mystery.") Later still, grasses and cacti evolved.

Through it all, plants continued to make a living more or less the same way they had since that ancient cyanobacterium took up with the alga. Photosynthesis remained remarkably stable over thousands of millennia of natural selection. It didn't change when humans began to domesticate plants, ten thousand years ago, or, later, when they figured out how to irrigate, fertilize, and, finally, hybridize them. It always worked well enough to power the planet—that is, until now.

STEPHEN LONG IS a professor of plant biology and crop sciences at the University of Illinois Urbana-Champaign and the director of a project called Realizing Increased Photosynthetic Efficiency, or RIPE. The premise of RIPE is that, as remarkable as photosynthesis may be, it needs to do better.

At seventy-one, Long is thin and fit, with a craggy face and a voice so soft it borders on a murmur. He grew up in London in a working-class family and attended what he describes as "not the best" high school. (It's since been closed.) One of the teachers at the school stood out—a plant enthusiast who took her students on frequent field trips. Inspired, Long decided to study agricultural botany at the University of Reading. Midway to his degree, he took a year off to work for a British food company, Tate & Lyle, which owned sugarcane plantations in the Caribbean and did a lot of sugar refining. Some at the company thought it might be possible to dispense with the plantations and even the cane and coax plant cells to produce sugar in vats. The idea didn't pan out—"It never became economically feasible," Long told me when, in July, I went to visit him at his office—but it got him interested in the mechanics of photosynthesis.

Photosynthesis takes place within a plant's chloroplasts—tiny organelles that are the descendants of that original captured cyanobacterium. When a photon is absorbed by a chloroplast, it initiates a cascade of reactions that convert light into chemical energy. These reactions are mediated by proteins, which are encoded by genes. Through a second series of reactions, the chemical energy is used to build carbohydrates. This requires more proteins. Photosynthesis has been called "one of the most complex of all biological processes," and when Long was starting out a great deal was still unknown about how, exactly, it worked. Gradually, using new molecular tools, researchers succeeded in filling in the gaps. Photosynthesis, they learned, requires the completion of some hundred and fifty discrete steps and involves roughly that number of genes.

The more that was discovered about the intricacies of photosynthesis, the more was revealed about its inefficiency. The comparison is often made to photovoltaic cells. Those on the market today convert about 20 percent of the sunlight that strikes them into electricity, and, in labs, researchers have achieved rates of almost 50 percent. Plants convert only about 1 percent of the sunlight that hits them into growth. In the case of crop plants, on average only about half of 1 percent of the light is converted into energy that people can use. The contrast isn't really fair to biology, since plants construct themselves, whereas PV cells have to be manufactured with energy from another source. Plants also store their own energy, while PV cells require separate batteries for that. Still, researchers who have tried to make apples-to-apples (or silicon-to-carbon) calculations have concluded that plants come out the losers.

Long went on to get a PhD and then took a teaching job at the University of Essex, on England's east coast. He became convinced that photosynthesis's inefficiency presented an opportunity. If the process could be streamlined, plants that had spent millennia just chugging along could become champions. For agriculture, the implications were profound. Potentially, new crop varieties could be created that could produce more with less.

"All of our food, directly or indirectly, comes from the process of photosynthesis," Long told me. "And we know that even our very best crops are only achieving a fraction of photosynthesis's theoretical efficiency. So, if we can work out how to improve photosynthesis, we can boost yields. We won't have to go on destroying yet more land for crops—we can try to produce more on the land we're already using."

Other biologists were skeptical. Surely, they observed, if there were a way to improve photosynthesis that was truly viable, and not just theoretical, then, at some point during the past several hundred million years, plants would have hit upon it. What their argument missed, Long thought, were the exigencies of evolution itself. To be preserved, biological systems don't have to be optimized. They just have to be functional.

"Evolution is not really about being productive," Long told me. "It's about getting your genes into the next generation."

IN 1999, Long decided that he would create his own version of photosynthesis. By this time, he'd moved to the University of Illinois, where many of the major discoveries about the process had been made. Long's idea was to build a computer simulation that would model each of the hundred-and-fifty-odd steps in photosynthesis as a differential equation. The effort dragged on for years, in part because Long's program kept crashing. Eventually, he got in touch with a computer scientist who worked for NASA on rocket engines.

"He said, 'Oh, I had exactly the same problem, and this is the routine I used,'" Long recalled. "And we worked with him and used that routine, and, bingo, it worked." Because photosynthesis is so complicated, and because the math involved is also complicated, Long's model requires a phenomenal amount of computing power. To simulate the performance of a single leaf over the course of a few minutes, it must make millions of calculations.

Once his model, which he dubbed e-photosynthesis, was up and running, Long could create new leaves without the bother of actually growing anything. He could probe the weaknesses of photosynthesis and test possible fixes. What would happen, for example, if a certain gene were ginned up to produce more of a certain enzyme? Would this accelerate photosynthesis or just gum up the works? The model would analyze the results of each virtual intervention, or hack. "Of course, ninety-nine times out of a hundred you're making things worse," Long said.

It was the hundredth hack that kept things interesting. Long found that, by rejiggering certain steps, nature could be improved upon. In 2006, he published a paper outlining half a dozen "opportunities for increasing photosynthesis." Among the people intrigued by the idea were some high-level staff members at the Bill and Melinda Gates Foundation. In 2011, the foundation invited Long and some of his colleagues to Seattle to discuss their work. Six months later, the foundation invited the group back. Long and his collaborators spent a week on Bainbridge Island, in Puget Sound, drawing up a funding proposal, and on the last day of their stay they presented their pitch to Bill Gates. In 2012, the foundation awarded them twenty-five million dollars, and RIPE was created. Later, the project received additional funding from Britain's Foreign, Commonwealth, and Development Office and from the Foundation for Food and Agriculture, a joint public-private venture based in Washington, D.C.

"It will take multiple innovations to solve the global food crisis," Gates told me via email. These include seed varieties that can better withstand drought, crops that can better fight off disease, and "game-changing discoveries that will lead to better harvests."

One of the opportunities that Long identified in his 2006 paper involves a process known as nonphotochemical quenching, or NPQ. Obviously, plants need light, but, like us, they can suffer from too much of it. NPQ enables them to protect themselves by dissipating excess light as heat. The problem is that NPQ is sluggish; once initiated, it's slow to stop, even as light conditions change. Long's model suggested that some clever genetic modifications could make the process nimbler.

Researchers at RIPE set about testing this proposition on tobacco plants, which are sort of the lab rats of the ag world. They inserted three extra genes into the plants, then raised them in greenhouses. The modified plants did, indeed, outperform ordinary tobacco plants—they grew faster and put on more weight. The team then ran field trials. Long nervously awaited the outcome. The results were even better than he'd hoped: the modified plants outperformed the control plants by up to 20 percent.

When the resulting paper was published, in *Science,* it made news around the world. "Genetic breakthrough," the BBC declared. Long was interviewed by the Big Ten Network, which, in addition to airing the conference's sporting events, sometimes does features on Big Ten professors. He told the interviewer that the day the results of the field trials came in was one of the most exciting of his life. "Don't tell my wife that," he added. The network showed the clip on the jumbotron during a University of Illinois football game. Long and his wife, Ann, were watching at home.

"I got an elbow in the ribs for that," he recalled.

IN 1967, two sober-minded men published a book with a sensational title: *Famine—1975!* The authors, William and Paul Paddock, were brothers; William was an agronomist, Paul a retired Foreign Service officer. "A collision between exploding population and static agriculture is imminent," the Paddocks wrote. They declared, "The conclusion is clear: there is no possibility of improving agriculture . . . soon enough to avert famine."

Many experts shared their anxiety. In the mid-sixties, the global population was growing by more than 2 percent a year, which is believed to be the highest rate in human history. In a number of developing countries—Brazil and Ethiopia, for instance—the annual rate was closer to 3 percent. Agricultural production wasn't keeping up.

"The world food situation is now more precarious than at any time since the period of acute shortage immediately after the Second World War," the director-general of the United Nations Food and Agriculture Organization, Binay Ranjan Sen, wrote. He warned that unless dramatic action was taken "Malthusian correctives" would "inexorably come into play."

Famine—1975! was followed by *The Population Bomb,* by the Stanford biologist Paul Ehrlich, published in 1968. Ehrlich, too, declared disaster unavoidable. "The battle to feed all of humanity is over," he wrote. "In the 1970's the world will undergo famines—hundreds of millions of people are going to starve to death in spite of any crash programs embarked upon now." Ehrlich became a regular guest on *The Tonight Show,* and *The Population Bomb* sold more than two million copies.

The catastrophe failed to materialize. Ehrlich and the Paddocks were wrong about the future of agriculture. Even as they were writing, the seeds—both literal and metaphorical—were being sown for what would become known as the Green Revolution.

At the vanguard of the revolution was Norman Borlaug, a plant pathologist who worked for the Rockefeller Foundation at an agricultural research station in Mexico. By painstakingly breeding wheat over the course of two decades, he developed a series of highly productive, disease-resistant varieties. The varieties were unusually stocky—they'd been bred using dwarf strains—and this allowed them to put more energy into their kernels and less into their stalks. As the varieties were adopted, yields shot up; in the two decades following the publication of *Famine—1975!,* wheat production in Mexico nearly doubled. During the same period in India, it more than tripled.

Building on Borlaug's work, breeders in the Philippines created high-yield, semi-dwarf strains of rice, which led to similar productivity increases. This work was motivated as much by political impulses as by humanitarian ones; boosting rice output might be described as the "hearts and bellies" approach to fighting communism in Asia.

For his efforts, Borlaug was awarded the Nobel Peace Prize in 1970. "More than any other single person of this age, he has helped to provide bread for a hungry world," the chairwoman of the Norwegian Nobel Committee stated.

Like most revolutions, the green one had unintended consequences. The new, high-yield varieties were needy; to realize their full potential, they required plenty of fertilizer, pesticides, and water. These "inputs," in turn, required money. The bulk of the benefits thus accrued to those with resources. Farms became bigger and more mechanized, developments that often cost the very poorest agricultural workers their livelihoods. Research suggests that the new varieties, combined with the agricultural practices they promoted, exacerbated inequality.

"The availability of 60% cheaper rice would be little consolation to someone who had lost 100% of their income as a result of the Green Revolution," Raj Patel, a research professor at the University of Texas at Austin, has written.

The ecological costs, too, were high, and by many accounts these are still growing. Fertilizer runoff has filled rivers and lakes with nutrients, producing algae blooms and aquatic "dead zones." Increased pesticide use has had the perverse effect of doing in many of the beneficial insects that once kept pests in check. The demands of irrigation have emptied aquifers. In the northern Indian state of Punjab, an early center of the Green Revolution, groundwater is being pumped out so much faster than it can be replenished that the water table is falling by about three feet a year. Experts have warned that, if current rates of pumping continue, in twenty-five years the state, which is sometimes referred to as "the food bowl of India," could be reduced to a desert.

"The situation is alarming," Rana Gurjit Singh, a member of Punjab's Legislative Assembly, observed a few months ago. "It is time to wake up."

It is often said that the world now needs a New Green Revolution, or a Second Green Revolution, or Green Revolution 2.0. The rate of yield growth for crops like wheat, rice, and corn appears to be plateauing, and the number of people who are hungry is once again on the rise. The world's population, meanwhile, continues to increase; now almost eight billion, it's projected to reach nearly ten billion by 2050. Income gains in countries like China are increasing the consumption of meat, which requires ever more grain and forage to produce. To meet the expected demand, global agricultural output will have to rise by almost 70 percent during the next thirty years. Such an increase would be tough to achieve in the best of times, which the coming decades are not likely to be. Recent research suggests that climate change has already begun to cut into yields, and, as the planet warms, the bite will only get bigger. (Agriculture itself is a major contributor to climate change.) Devoting more land to farming isn't really an option, or, at least, not a good one. Most of the world's best soils are already under cultivation, and mowing down forests to plant corn or soybeans would lead to still more warming.

"At no other point in history has agriculture been faced with such an array of familiar and unfamiliar risks" is how a recent report from the Food and Agriculture Organization put it.

"We need to up our game," Enock Chikava, who grew up on a ten-acre farm in Zimbabwe and now serves as the interim director for agricultural development at the Gates Foundation, told me. "We can't continue business as usual."

ONE DAY WHILE I was in Urbana, Long took me to visit RIPE's test fields. This was in the midst of one of last summer's brutal heat waves, and to avoid the midmorning sun we met up at eight a.m. Even so, it was sweltering.

RIPE's test plots are to the average farm what a Tesla is to a Model T. Looming above the plots are hundred-and-fifty-foot-tall metal towers strung with guy wires. The wires are controlled by computerized winches imported from Austria—a setup that was originally devised to film professional sports matches. RIPE's setup carries sensors that, among other things, shoot out laser beams and detect infrared radiation. When I visited, the sensors had just been installed; the idea was to track the plants' progress on a day-to-day basis.

Long led me over to a plot surrounded by an electric fence. It was divided into forty identical rectangles, each studded with white tags. The rectangles were planted with different strains of genetically modified soybeans, which had been tweaked in much the same way that the tobacco plants had, to speed up NPQ. Long bent over some rows labeled E27.

"I might be imagining, but it looks like these are a little bit taller," he said. He quickly added, "You've got to be very careful at this stage, though." In the summer of 2020, the tweaked soy plants had produced significantly more soybeans than the control ones did. E27 had performed particularly well. But was this just a fluke? "We're hoping to get the definitive answer this year," Long told me.

In another plot, tobacco plants were growing low to the ground. These, he explained, represented an effort to address a different drag on photosynthesis, involving the enzyme RuBisCo.

To make sugars, plants use carbon dioxide they've taken in from the air. RuBisCo, which is believed to be the most abundant enzyme on the planet, in effect grabs the CO_2 and sends it on to the sugar-making process. Like NPQ, RuBisCo is slow. Even more significantly, it's error-prone. Sometimes, like an assembly-line worker who picks up the wrong part, it grabs a molecule of oxygen instead of carbon dioxide. (Presumably, RuBisCo makes this mistake because at the point it was first synthesized, billions of years ago, there was hardly any oxygen around to worry about.) When RuBisCo accidentally picks up O_2, the plant produces a compound that's toxic, which it then has to get rid of. The exercise is quite costly: it's estimated that it can reduce the efficiency of photosynthesis by 40 percent. Using genes from bacteria and algae, the RIPE team has developed "bypass" tobacco plants, which break down the toxic compound in fewer steps.

Long pointed to a muddy plot nearby. Had I arrived a few weeks earlier, he said, I would have found "bypass" potatoes growing there. These had been destroyed by heavy rains, and now it was too late in the season to replant. "It's kind of been wrecked," he said, with a sigh.

From the fields, we drove to an enormous greenhouse. Before entering it, we had to put on lab coats and sterile booties. Near the door were benches of tobacco plants wrapped in cellophane. The rest of the greenhouse was filled with long rows of what looked like DVD players. These turned out to be high-tech scales connected to a precision irrigation system. Plants could be placed on the scales and given measured sips of water; then they'd be automatically weighed to see how much bulk they'd put on. More than four hundred plants could be tested at once, and the results would quickly reveal which specimens with which genetic changes were the best performers. Someone flipped a switch, and a set of cameras mounted on scaffolding began to creep over the rows. The cameras, I was told, would produce a continuous stream of data about the plants, so that everything down to the curve of their leaves could be studied.

Since its founding, in 2012, RIPE has expanded to include almost a hundred researchers across four continents. Long's hope is that, in addition to the NPQ and bypass tweaks, the project will come up with half a dozen other ways to "improve" photosynthesis. A team in Australia is looking at how to speed carbon dioxide's journey to RuBisCo, and a team in England is looking at what happens right after RuBisCo does its job. The next step would be to get these genetic modifications into globally significant crop plants—in addition to soy and potatoes, RIPE is working with corn, cowpeas, and cassava—and then into local varieties. (Farmers in different parts of the world plant different strains of corn and cassava that have been bred for local conditions.)

Long is particularly keen on getting photosynthetically souped-up seed to farmers in sub-Saharan Africa, a region that didn't much benefit from the yield gains of the original Green Revolution. Today, more than two hundred million people there are chronically undernourished.

"If we can provide smallholder farmers in Africa with technologies that will produce more food and give them a better livelihood, that's what really motivates the team," Long told me. One of the Gates Foundation's stipulations is that any breakthroughs that result from RIPE's work be made available "at an affordable price" to companies or government agencies that supply seed to farmers in the world's poorest countries.

Before any of RIPE's creations could be planted in sub-Saharan Africa, though, or anywhere else, for that matter, all sorts of licenses would have to be obtained. (The gene-editing techniques that Long and his colleagues are using are themselves often patented.) Then the altered genes would have to be approved by the relevant agency in the nation in question, and the alterations would have to be bred into local varieties. So far, only a handful of African countries have okayed genetically modified crops, and most of the approvals have been for GM cotton. A recent study noted that at least two dozen GM food crops—some modified for insect resistance, others for salt tolerance—have been submitted to regulatory agencies in the region but remain in limbo.

"A host of viable technologies continue to sit on the shelf, frequently due to regulatory paralysis," the study observed. (In the United States, practically all of the soy and corn grown is genetically modified; other approved GM food crops include apples, potatoes, papayas, sugar beets, and canola. In Europe, by contrast, GM crops are generally banned.) Meanwhile, to the extent that attitudes toward GM foods have been surveyed in sub-Saharan Africa, a majority of people seem to be leery of them. A recent study conducted in Zimbabwe, for example, found that almost three-quarters of the respondents believed them to be "too risky." And smallholder farmers don't have enough land to leave buffer zones, which means that, if they grow GM crops that cross-pollinate, these could mix with, or contaminate, their non-GM neighbors.

When I asked Long about the advisability of developing genetically modified varieties for use in countries that don't particularly seem to want them, he told me that, at a meeting with RIPE researchers, a similar question had been posed to Bill Gates.

"His response was 'Well, things might change if these predictions of food shortages come to pass,'" Long said. "'And, if they do come to pass, it's going to be too late to do this research.'"

SOME THIRTY MILLION years ago, a plant—no one knows exactly which one, but probably it was a grass—came up with its own hack to improve photosynthesis. The hack didn't alter the steps involved in the process; instead, it added new ones. The new steps concentrated CO_2 around RuBisCo, effectively eliminating the enzyme's opportunity to make a mistake. (To extend the assembly-line metaphor, imagine a worker surrounded by crateloads of the right parts and none of the wrong ones.) At the time, carbon dioxide levels in the atmosphere were falling—a trend that would continue more or less until humans figured out how to burn fossil fuels—so even though the hack cost the plant some energy, it offered a net gain. In fact, it proved so useful that other plants soon followed suit. What's now known as C4 photosynthesis evolved independently at least forty-five times, in nineteen different plant families. (The term *C4* refers to a four-carbon compound that's produced in one of the supplemental steps.) Nowadays, several of the world's key crop plants are C4, including corn, millet, and sorghum, and so are several of the world's key weeds, like crabgrass and tumbleweed.

C4 photosynthesis isn't just more efficient than ordinary photosynthesis, which is known as C3. It also requires less water and less nitrogen, and so, in turn, less fertilizer. About twenty-five years ago, a plant physiologist named John Sheehy came up with what many other plant physiologists considered to be an absurd idea. He decided that rice, which is a C3 plant, should be transformed into a C4. Like Long, Sheehy was from England, but he was working in the Philippines, at the research institute where, in the nineteen-sixties, breeders had developed the rice varieties that helped spark the Green Revolution. In 1999, Sheehy hosted a meeting at the institute to discuss his idea. The general opinion of the participants was that it was impossible.

Sheehy didn't give up. In 2006, nearing retirement, he pulled together a second meeting on the topic. Again, the attendees were skeptical. But this time around they decided that Sheehy's scheme was at least worth a try. Jane Langdale, a plant biologist from Oxford, was among the researchers at the second meeting. "There was a sense that it was now or never," she said recently, when I spoke to her over Zoom. "We were either going to have to get younger people interested in this or lose the opportunity." Thus was born the C4 Rice Project, which Langdale now heads. (Sheehy died in 2019.)

The C4 Rice Project could be thought of as RIPE's edgier cousin. It, too, is funded by the Gates Foundation, and it, too, aims to feed the world by reengineering it from the chloroplast up. "Given that the C4 pathway is up to 50% more efficient than the C3 pathway, introducing C4 traits into a C3 crop would have a dramatic impact on crop yield," the project's website observes.

What makes the work so challenging is that C4 plants don't just go through extra steps in photosynthesis; they have a different anatomy. Among other things, the veins in the leaves of C4 plants are much more closely packed than those in C3 plants, and this spacing is crucial to the enterprise. The C4 Rice Project involves thirty researchers in five countries. Some of the scientists are focused on transforming the plant's leaves, others on altering its biochemistry.

"We're working to try to do these two things in parallel," Langdale explained to me. "But ultimately we have to do them both."

The project has run into lots of obstacles; still, it has inched forward. Langdale's lab has succeeded in producing rice plants with a greater volume of veins in their leaves, though the volume is still not quite high enough. Other labs have developed rice plants that generate the crucial four-carbon compound; these plants, however, don't take the next step, which is to give up one of the carbons to be grabbed by RuBisCo.

"When we started, everybody thought we were mad," Langdale said. "And it has not been an easy journey. But I think now people look and think, 'You know—they actually are making progress.'

"I don't know whether we'll ever make rice with the full C4 anatomy and the biochemistry," she continued. "But I do think along the way we are going to find things that improve yield and improve efficiency, even if it's not the full shebang."

A FEW DAYS after I spoke to Langdale, three Punjabi villagers were hit by a truck at the site of a demonstration near New Delhi. (The victims were all women in their fifties and sixties.) During the past year, hundreds of thousands of farmers in India have protested against the government of Prime Minister Narendra Modi, and for months tens of thousands have been camped out along the roads leading into the capital.

In an immediate sense, the target of the farmers' ire is a set of laws pushed through Parliament by Modi's party; these, they fear, could lead to an end to government price supports. In a deeper sense, though, the tensions go back to the Green Revolution. To encourage farmers to plant the higher-yielding, thirstier varieties of rice and wheat, the Indian government introduced the price-support system in the nineteen-sixties. Now the subsidies have produced gluts of these commodities, even as growing them is depleting the country's aquifers, and the government wants to prod farmers to move away from the crops it once prodded them to plant. To the country's millions of farmers, most of whom own fewer than five acres, changes in the status quo seem likely to lead only to more misery.

"Many people would argue that the price supports that are currently given are barely adequate to cover the costs of production," Sudha Narayanan, a research fellow at the International Food Policy Research Institute's office in New Delhi, told me. But farmers depend on the supports to at least set a floor on their incomes: "They are seen as a kind of insurance." Late last month, in a surprise move, Parliament voted to repeal the laws, but that has not put an end to the protests; farmers are now calling for an extension of price supports to other crops.

How to produce a second Green Revolution without repeating, or compounding, the mistakes of the first is a question that dogs efforts to boost yields, particularly in the Global South. With climate change, the challenges are, in many ways, even steeper than they were in the nineteen-sixties. The research institutes that helped drive the original Green Revolution, which include the International Maize and Wheat Improvement Center, in Mexico, where Norman Borlaug was stationed, and the International Rice Research Institute, in the Philippines, where John Sheehy worked, are part of a consortium called CGIAR (Consultative Group on International Agricultural Research). CGIAR is in the midst of restructuring itself.

"Fundamentally, the reorganization is about trying to attack what we call twenty-first-century problems, paying attention to the critique of the Green Revolution," Channing Arndt, a division director at the International Food Policy Research Institute, which is part of CGIAR, told me. The Green Revolution "definitely brought a lot of calories," he continued. "But it also brought pollution and other problems, which we don't want to repeat."

One way to look at RIPE and the C4 Rice Project is as efforts to bring twenty-first-century tools to bear on twenty-first-century problems. For better or worse, we now have the ability to tinker with life at the most basic level, and this opens up all sorts of possibilities, from treating genetic disorders to manufacturing biological weapons. Crop plants that make fewer mistakes in photosynthesis, or that complete the process more efficiently, would produce more food per acre, potentially with fewer inputs. Not only humans would benefit; so, too, would the myriad species whose habitats would be spared. "Twenty years from now, this could be making a major difference," Edward Mabaya, a research professor at Cornell, told me.

But, in many ways, the twenty-first century's problems are hold-overs from the nineteenth and twentieth centuries, and it's not clear whether the new tools are a better match for them than the old. As Mabaya, who also serves as the chief scientific adviser for the African Seed Access Index, pointed out to me, researchers have already developed plenty of improved varieties for sub-Saharan Africa, using conventional breeding methods.

"Most of the varieties, maybe 80 percent of them, just end up on the shelf," he said. "They never reach smallholder farmers." (The Access Index, which is working to identify the choke points in African seed systems, is another group funded, in part, by the Gates Foundation.)

Vara Prasad, a crop scientist at Kansas State University and the director of one of its Feed the Future Innovation Labs, made much the same point to me: a majority of the smallholder farmers in Africa and South Asia aren't planting the improved varieties that already exist. Sometimes the issue is cost. For instance, with hybrids, the seeds can't be saved and have to be repurchased every year; though the extra yield should cover the expense, smallholder farmers may just not have the cash. Sometimes the obstacles can be difficult even to identify.

"We always talk about the technologies, but we ignore the social piece," Prasad told me. "We need to understand the barriers to adoption, and we don't have a clear understanding of those.

"I've looked at the RIPE project," he went on. "Are there anthropologists on it? Any economists? Any nutrition folks? Gender-empowerment folks? We really need to be thinking about social innovation here, not only biophysical innovation—and I'm a biophysical scientist."

Borlaug himself warned against putting too much faith in technology to solve society's ills. In his Nobel Lecture, in 1970, he called the Green Revolution a "temporary success"; if the population continued to climb, this success, he feared, would prove "ephemeral."

"There are no miracles in agricultural production," he said. And, even if production could keep up with population growth, there would remain the issue of distribution, of bridging the great global divide between the haves, who "live in a luxury never before experienced," and the have-nots, who send their kids to bed hungry.

"It is a sad fact that on this earth at this late date there are still two worlds," Borlaug observed.

Originally published in *The New Yorker,*
December 13, 2021.

GOING NEGATIVE

Can Carbon Dioxide Removal
Save the World?

CARBON ENGINEERING, a company owned in part by Bill Gates, has its headquarters on a spit of land that juts into Howe Sound, an hour north of Vancouver. Until recently, the land was a toxic waste site, and the company's equipment occupies a long, barn-like building that, for many years, was used to process contaminated water. The offices, inherited from the business that poisoned the site, provide a spectacular view of Mt. Garibaldi, which rises to a snow-covered point, and of the Chief, a granite monolith that's British Columbia's answer to El Capitan. To protect the spit against rising sea levels, the local government is planning to cover it with a layer of fill six feet deep. When that's done, it's hoping to sell the site for luxury condos.

Adrian Corless, Carbon Engineering's chief executive, who is fifty-one, is a compact man with dark hair, a square jaw, and a concerned expression. "Do you wear contacts?" he asked, as we were suiting up to enter the barnlike building. If so, I'd have to take extra precautions, because some of the chemicals used in the building could cause the lenses to liquefy and fuse to my eyes.

Inside, pipes snaked along the walls and overhead. The thrum of machinery made it hard to hear. In one corner, what looked like over-sized beach bags were filled with what looked like white sand. This, Corless explained over the noise, was limestone—pellets of pure calcium carbonate.

Corless and his team are engaged in a project that falls somewhere between toxic waste cleanup and alchemy. They've devised a process that allows them, in effect, to suck carbon dioxide out of the air. Every day at the plant, roughly a ton of CO_2 that had previously floated over Mt. Garibaldi or the Chief is converted into calcium carbonate. The pellets are subsequently heated, and the gas is forced off, to be stored in cannisters. The calcium can then be recovered, and the process run through all over again.

"If we're successful at building a business around carbon removal, these are trillion-dollar markets," Corless told me.

This past April, the concentration of carbon dioxide in the atmosphere reached a record four hundred and ten parts per million. The amount of CO_2 in the air now is probably greater than it's been at any time since the mid-Pliocene, three and a half million years ago, when there was a lot less ice at the poles and sea levels were sixty feet higher. This year's record will be surpassed next year, and next year's the year after that. Even if every country fulfills the pledges made in the Paris climate accord—and the United States has said that it doesn't intend to—carbon dioxide could soon reach levels that, it's widely agreed, will lead to catastrophe, assuming it hasn't already done so.

Carbon dioxide removal is, potentially, a trillion-dollar enterprise because it offers a way not just to slow the rise in CO_2 but to reverse it. The process is sometimes referred to as "negative emissions": instead of adding carbon to the air, it subtracts it. Carbon removal plants could be built anywhere, or everywhere. Construct enough of them and, in theory at least, CO_2 emissions could continue unabated and still we could avert calamity. Depending on how you look at things, the technology represents either the ultimate insurance policy or the ultimate moral hazard.

CARBON ENGINEERING IS one of a half-dozen companies vying to prove that carbon removal is feasible. Others include Global Thermostat, which is based in New York, and Climeworks, based near Zurich. Most of these owe their origins to the ideas of a physicist named

Klaus Lackner, who now works at Arizona State University, in Tempe, so on my way home from British Columbia I took a detour to visit him. It was July, and on the day I arrived the temperature in the city reached a hundred and twelve degrees. When I got to my hotel, one of the first things I noticed was a dead starling lying, feet up, in the parking lot. I wondered if it had died from heat exhaustion.

Lackner, who is sixty-five, grew up in Germany. He is tall and lanky, with a fringe of gray hair and a prominent forehead. I met him in his office at an institute he runs, the Center for Negative Carbon Emissions. The office was bare, except for a few *New Yorker* cartoons on the theme of nerd-dom, which, Lackner told me, his wife had cut out for him. In one, a couple of scientists stand in front of an enormous whiteboard covered in equations. "The math is right," one of them says. "It's just in poor taste."

In the late nineteen-seventies, Lackner moved from Germany to California to study with George Zweig, one of the discoverers of quarks. A few years later, he got a job at Los Alamos National Laboratory. There, he worked on fusion. "Some of the work was classified," he said, "some of it not."

Fusion is the process that powers the stars and, closer to home, thermonuclear bombs. When Lackner was at Los Alamos, it was being touted as a solution to the world's energy problem; if fusion could be harnessed, it could generate vast amounts of carbon-free power using isotopes of hydrogen. Lackner became convinced that a fusion reactor was, at a minimum, decades away. (Decades later, it's generally agreed that a workable reactor is still decades away.) Meanwhile, the globe's growing population would demand more and more energy, and this demand would be met, for the most part, with fossil fuels.

"I realized, probably earlier than most, that the claims of the demise of fossil fuels were greatly exaggerated," Lackner told me. (In fact, fossil fuels currently provide about 80 percent of the world's energy. Proportionally, this figure hasn't changed much since the mid-eighties, but, because global energy use has nearly doubled, the amount of coal, oil, and natural gas being burned today is almost two times greater.)

One evening in the early nineties, Lackner was having a beer with a friend, Christopher Wendt, also a physicist. The two got to wondering why, as Lackner put it to me, "nobody's doing these really crazy, big things anymore." This led to more questions and more conversations (and possibly more beers).

Eventually, the two produced an equation-dense paper in which they argued that self-replicating machines could solve the world's energy problem and, more or less at the same time, clean up the mess humans have made by burning fossil fuels. The machines would be powered by solar panels, and as they multiplied they'd produce more solar panels, which they'd assemble using elements, like silicon and aluminum, extracted from ordinary dirt. The expanding collection of panels would produce ever more power, at a rate that would increase exponentially. An array covering three hundred and eighty-six thousand square miles—an area larger than Nigeria but, as Lackner and Wendt noted, "smaller than many deserts"—could supply all the world's electricity many times over.

This same array could be put to use scrubbing carbon dioxide from the atmosphere. According to Lackner and Wendt, the power generated by a Nigeria-size solar farm would be enough to remove all the CO_2 emitted by humans up to that point within five years. Ideally, the CO_2 would be converted to rock, similar to the white sand produced by Carbon Engineering; enough would be created to cover Venezuela in a layer a foot and a half deep. (Where this rock would go the two did not specify.)

Lackner let the idea of the self-replicating machine slide, but he became more and more intrigued by carbon dioxide removal, particularly by what's become known as "direct air capture."

"Sometimes by thinking through this extreme end point you learn a lot," he said. He began giving talks and writing papers on the subject. Some scientists decided he was nuts, others that he was a visionary. "Klaus is, in fact, a genius," Julio Friedmann, a former principal deputy assistant secretary of energy and an expert on carbon management, told me.

In 2000, Lackner received a job offer from Columbia University. Once in New York, he pitched a plan for developing a carbon-sucking technology to Gary Comer, a founder of Lands' End. Comer brought to the meeting his investment adviser, who quipped that Lackner wasn't looking for venture capital so much as "adventure capital." Nevertheless, Comer offered to put up five million dollars. The new company was called Global Research Technologies, or GRT. It got as far as building a small prototype, but just as it was looking for new investors the financial crisis hit.

"Our timing was exquisite," Lackner told me. Unable to raise more funds, the company ceased operations. As the planet continued to warm, and carbon dioxide levels continued to climb, Lackner came to believe that, unwittingly, humanity had already committed itself to negative emissions.

"I think that we're in a very uncomfortable situation," he said. "I would argue that if technologies to pull CO_2 out of the environment fail then we're in deep trouble."

LACKNER FOUNDED THE Center for Negative Carbon Emissions at ASU in 2014. Most of the equipment he dreams up is put together in a workshop a few blocks from his office. The day I was there, it was so hot outside that even the five-minute walk to the workshop required staging. Lackner delivered a short lecture on the dangers of dehydration and handed me a bottle of water.

In the workshop, an engineer was tinkering with what looked like the guts of a foldout couch. Where, in the living-room version, there would have been a mattress, in this one was an elaborate array of plastic ribbons. Embedded in each ribbon was a powder made from thousands upon thousands of tiny, amber-colored beads. The beads, Lackner explained, could be purchased by the truckload; they were composed of a resin normally used in water treatment to remove chemicals like nitrates. More or less by accident, Lackner had discovered that the beads could be repurposed. Dry, they'd absorb carbon dioxide. Wet, they'd release it. The idea was to expose the ribbons to Arizona's thirsty air and then fold the device into a sealed container filled with water. The CO_2 that had been captured by the powder in the dry phase would be released in the wet phase; it could then be piped out of the container, and the whole process restarted, the couch folding and unfolding over and over again.

Lackner has calculated that an apparatus the size of a semitrailer could remove a ton of carbon dioxide per day, or three hundred and sixty-five tons a year. The world's cars, planes, refineries, and power plants now produce about thirty-six billion tons of CO_2 annually, so, he told me, "if you built a hundred million trailer-size units you could actually keep up with current emissions." He acknowledged that the figure sounded daunting. But, he noted, the iPhone has been around for only a decade or so, and there are now seven hundred million in use. "We are still very early in this game," he said.

The way Lackner sees things, the key to avoiding "deep trouble" is thinking differently. "We need to change the paradigm," he told me. Carbon dioxide should be regarded the same way we view other waste products, like sewage or garbage. We don't expect people to stop producing waste. ("Rewarding people for going to the bathroom less would be nonsensical," Lackner has observed.) At the same time, we don't let them shit on the sidewalk or toss their empty yogurt containers into the street.

"If I were to tell you that the garbage I'm dumping in front of your house is 20 percent less this year than it was last year, you would still think I'm doing something intolerable," Lackner said.

One of the reasons we've made so little progress on climate change, he contends, is that the issue has acquired an ethical charge, which has polarized people. To the extent that emissions are seen as bad, emitters become guilty. "Such a moral stance makes virtually everyone a sinner, and makes hypocrites out of many who are concerned about climate change but still partake in the benefits of modernity," he has written. Changing the paradigm, Lackner believes, will change the conversation. If CO_2 is treated as just another form of waste, which has to be disposed of, then people can stop arguing about whether it's a problem and finally start doing something.

• • •

CARBON DIOXIDE WAS "discovered," by a Scottish physician named Joseph Black, in 1745. A decade later, another Scotsman, James Watt, invented a more efficient steam engine, ushering in what is now called the age of industrialization but that future generations may dub the age of emissions. It is likely that by the end of the nineteenth century human activity had raised the average temperature of the earth by a tenth of a degree Celsius (or nearly two-tenths of a degree Fahrenheit).

As the world warmed, it started to change, first gradually and then suddenly. By now, the globe is at least one degree Celsius (1.8 degrees Fahrenheit) warmer than it was in Black's day, and the consequences are becoming ever more apparent. Heat waves are hotter, rainstorms more intense, and droughts drier. The wildfire season is growing longer, and fires, like the ones that recently ravaged Northern California, more numerous. Sea levels are rising, and the rate of rise is accelerating. Higher sea levels exacerbated the damage from Hurricanes Harvey, Irma, and Maria, and higher water temperatures probably also made the storms more ferocious. "Harvey is what climate change looks like," Eric Holthaus, a meteorologist turned columnist, recently wrote.

Meanwhile, still more warming is locked in. There's so much inertia in the climate system, which is as vast as the earth itself, that the globe has yet to fully adjust to the hundreds of billions of tons of carbon dioxide that have been added to the atmosphere in the past few decades. Last month, the World Meteorological Organization announced that the concentration of carbon dioxide in the atmosphere jumped by a record amount in 2016.

No one can say exactly how warm the world can get before disaster— the inundation of low-lying cities, say, or the collapse of crucial ecosystems, like coral reefs—becomes inevitable. Officially, the threshold is two degrees Celsius (3.6 degrees Fahrenheit) above preindustrial levels. Virtually every nation signed on to this figure at a round of climate negotiations held in Cancun in 2010.

Meeting in Paris in 2015, world leaders decided that the two-degree threshold was too high; the stated aim of the climate accord is to hold "the increase in the global average temperature to well below 2°C" and to try to limit it to 1.5°C. Since the planet has already warmed by one degree and, for all practical purposes, is committed to another half a degree, it would seem impossible to meet the latter goal and nearly impossible to meet the former. And it *is* nearly impossible, unless the world switches course and instead of just adding CO_2 to the atmosphere also starts to remove it.

The extent to which the world is counting on negative emissions is documented by the latest report of the Intergovernmental Panel on Climate Change, which was published the year before Paris. To peer into the future, the IPCC relies on computer models that represent the world's energy and climate systems as a tangle of equations and that can be programmed to play out different "scenarios." Most of the scenarios involve temperature increases of two, three, or even four degrees Celsius—up to just over seven degrees Fahrenheit—by the end of this century. (In a recent paper in the *Proceedings of the National Academy of Sciences,* two climate scientists—Yang-yang Xu, of Texas A&M, and Veerabhadran Ramanathan, of the Scripps Institution of Oceanography—proposed that warming greater than three degrees Celsius be designated as "catastrophic" and warming greater than five degrees as "unknown??" The "unknown??" designation, they wrote, comes "with the understanding that changes of this magnitude, not experienced in the last 20+ million years, pose existential threats to a majority of the population.")

When the IPCC went looking for ways to hold the temperature increase under two degrees Celsius, it found the math punishing. Global emissions would have to fall rapidly and dramatically—pretty much down to zero by the middle of this century. (This would entail, among other things, replacing most of the world's power plants, revamping its agricultural systems, and eliminating gasoline-powered vehicles, all within the next few decades.) Alternatively, humanity could, in effect, go into hock. It could allow CO_2 levels temporarily to exceed the two-degree threshold—a situation that's become known as "overshoot"—and then, via negative emissions, pull the excess CO_2 out of the air.

The IPCC considered more than a thousand possible scenarios. Of these, only a hundred and sixteen limit warming to below two degrees, and of these a hundred and eight involve negative emissions. In many below-two-degree scenarios, the quantity of negative emissions called for reaches the same order of magnitude as the "positive" emissions being produced today.

"The volumes are outright crazy," Oliver Geden, the head of the EU research division of the German Institute for International and Security Affairs, told me. Lackner said, "I think what the IPCC really is saying is 'We tried lots and lots of scenarios, and, of the scenarios which stayed safe, virtually every one needed some magic touch of a negative emissions. If we didn't do that, we ran into a brick wall.'"

PURSUED ON THE scale envisioned by the IPCC, carbon dioxide removal would yield at first tens of billions and soon hundreds of billions of tons of CO_2, all of which would have to be dealt with. This represents its own supersized challenge. CO_2 can be combined with calcium to produce limestone, as it is in the process at Carbon Engineering (and in Lackner's self-replicating-machine scheme). But the necessary form of calcium isn't readily available, and producing it generally yields CO_2, a self-defeating prospect. An alternative is to shove the carbon back where it came from, deep underground.

"If you are storing CO_2, and your only purpose is storage, then you're looking for a package of certain types of rock," Sallie Greenberg, the associate director for energy, research, and development at the Illinois State Geological Survey, told me. It was a bright summer day, and we were driving through the cornfields of Illinois's midsection. A mile below us was a rock formation known as the Eau Claire Shale, and below that a formation known as the Mt. Simon Sandstone. Together with a team of drillers, engineers, and geoscientists, Greenberg has spent the past decade injecting carbon dioxide into this rock "package" and studying the outcome. When I'd proposed over the phone that she show me the project, in Decatur, she'd agreed, though not without hesitation.

"It isn't sexy," she'd warned me. "It's a wellhead."

Our first stop was a building shaped like a ski chalet. This was the National Sequestration Education Center, a joint venture of the Illinois Geological Survey, the U.S. Department of Energy, and Richland Community College. Inside were classrooms, occupied that morning by kids making lanyards, and displays aimed at illuminating the very dark world of carbon storage. One display was a sort of oversized barber pole, nine feet tall and decorated in bands of tan and brown, representing the various rock layers beneath us. A long arrow on the side of the pole indicated how many had been drilled through for Greenberg's carbon storage project; it pointed down, through the New Albany Shale, the Maquoketa Shale, and so on, all the way to the floor.

The center's director, David Larrick, was on hand to serve as a guide. In addition to schoolkids, he said, the center hosted lots of community groups, like Kiwanis clubs. "This is very effective as a visual," he told me, gesturing toward the pole. Sometimes farmers were concerned about the impact that the project could have on their water supply. The pole showed that the CO_2 was being injected more than a mile below their wells.

"We have had overwhelmingly positive support," he said. While Greenberg and Larrick chatted, I wandered off to play an educational video game. A cartoon figure in a hard hat appeared on the screen to offer factoids such as "The most efficient method of transport of CO_2 is by pipeline."

"Transport CO_2 to earn points!" the cartoon man exhorted.

After touring the center's garden, which featured grasses, like big bluestem, that would have been found in the area before it was plowed into cornfields, Greenberg and I drove on. Soon we passed through the gates of an enormous Archer Daniels Midland plant, which rose up out of the fields like a small city.

Greenberg explained that the project we were visiting was one of seven funded by the Department of Energy to learn whether carbon injected underground would stay there. In the earliest stage of the project, initiated under President George W. Bush, Greenberg and her colleagues sifted through geological records to find an appropriate test site. What they were seeking was similar to what oil drillers look for—porous stone capped by a layer of impermeable rock—only they were looking not to extract fossil fuels but, in a manner of speaking, to stuff them back in. The next step was locating a ready source of carbon dioxide. This is where ADM came in; the plant converts corn into ethanol, and one of the by-products of this process is almost pure CO_2. In a later stage of the project, during the Obama administration, a million tons of carbon dioxide from the plant were pumped underground. Rigorous monitoring has shown that, so far, the CO_2 has stayed put.

We stopped to pick up hard hats and went to see some of the monitoring equipment, which was being serviced by two engineers, Nick Malkewicz and Jim Kirksey. It was now lunchtime, so we made another detour, to a local barbecue place. Finally, Greenberg and I and the two men got to the injection site. It was, indeed, not sexy—just a bunch of pipes and valves sticking out of the dirt. I asked about the future of carbon storage.

"I think the technology's there and it's absolutely viable," Malkewicz said. "It's just a question of whether people want to do it or not. It's kind of an obvious thing."

"We know we can meet the objective of storing CO_2," Greenberg added. "Like Nick said, it's just a matter of whether or not as a society we're going to do it."

WHEN WORK BEGAN on the Decatur project, in 2003, few people besides Klaus Lackner were thinking about sucking CO_2 from the air. Instead, the goal was to demonstrate the feasibility of an only slightly less revolutionary technology—carbon capture and storage (or, as it is sometimes referred to, carbon capture and sequestration).

With CCS, the CO_2 produced at a power station or a steel mill or a cement plant is drawn off before it has a chance to disperse into the atmosphere. (This is called "postcombustion capture.") The gas, under very high pressure, is then injected into the appropriate package of rock, where it is supposed to remain permanently. The process has become popularly—and euphemistically—known as "clean coal," because, if all goes according to plan, a plant equipped with CCS produces only a fraction of the emissions of a conventional coal-fired plant.

Over the years, both Republicans and Democrats have touted clean coal as a way to save mining jobs and protect the environment. The coal industry has also, nominally at least, embraced the technology; one industry-sponsored group calls itself the American Coalition for Clean Coal Electricity. Donald Trump, too, has talked up clean coal, even if he doesn't seem to quite understand what the term means. "We're going to have clean coal, really clean coal," he said in March.

Currently, only one power plant in the United States, the Petra Nova plant, near Houston, uses postcombustion carbon capture on a large scale. Plans for other plants to showcase the technology have been scrapped, including, most recently, the Kemper County plant, in Mississippi. This past June, the plant's owner, Southern Company, announced that it was changing tacks. Instead of burning coal and capturing the carbon, the plant would burn natural gas and release the CO_2.

Experts I spoke to said that the main reason CCS hasn't caught on is that there's no inducement to use it. Capturing the CO_2 from a smokestack consumes a lot of power—up to 25 percent of the total produced at a typical coal-burning plant. And this, of course, translates into costs. What company is going to assume such costs when it can dump CO_2 into the air for free?

"If you're running a steel mill or a power plant and you're putting the CO_2 into the atmosphere, people might say, 'Why aren't you using carbon capture and storage?'" Howard Herzog, an engineer at MIT who for many years ran a research program on CCS, told me. "And you say, 'What's my financial incentive? No one's saying I can't put it in the atmosphere.' In fact, we've gone backwards in terms of sending signals that you're going to have to restrict it."

But, although CCS has stalled in practice, it has become ever more essential on paper. Practically all below-two-degree warming scenarios assume that it will be widely deployed. And even this isn't enough. To avoid catastrophe, most models rely on a yet-to-be-realized variation of CCS known as BECCS.

BECCS, which stands for "bioenergy with carbon capture and storage," takes advantage of the original form of carbon engineering: photosynthesis. Trees and grasses and shrubs, as they grow, soak up CO_2 from the air. (Replanting forests is a low-tech form of carbon removal.) Later, when the plants rot or are combusted, the carbon they have absorbed is released back into the atmosphere. If a power station were to burn wood, say, or cornstalks, and use CCS to sequester the resulting CO_2, this cycle would be broken. Carbon would be sucked from the air by the green plants and then forced underground. BECCS represents a way to generate negative emissions and, at the same time, electricity. The arrangement, at least as far as the models are concerned, could hardly be more convenient.

"BECCS is unique in that it removes carbon and produces energy," Glen Peters, a senior researcher at the Center for International Climate Research, in Oslo, told me. "So the more you consume the more you remove." He went on, "In a sense, it's a dream technology. It's solving one problem while solving the other problem. What more could you want?"

THE CENTER FOR Carbon Removal doesn't really have an office; it operates out of a co-working space in downtown Oakland. On the day I visited, not long after my trip to Decatur, someone had recently stopped at Trader Joe's, and much of the center's limited real estate was taken up by tubs of treats.

"Open anything you want," the center's executive director, Noah Deich, urged me, with a wave of his hand.

Deich, who is thirty-one, has a broad face, a brown beard, and a knowing sort of earnestness. After graduating from the University of Virginia, in 2009, he went to work for a consulting firm in Washington, D.C., that was advising power companies about how to prepare for a time when they'd no longer be able to release carbon into the atmosphere cost-free. It was the start of the Obama administration, and

that time seemed imminent. The House of Representatives had recently approved legislation to limit emissions. But the bill later died in
the Senate, and, as Deich put it, "It's no fun to model the impacts of
climate policies nobody believes are going to happen." He switched
consulting firms, then headed to business school at the University of
California, Berkeley.

"I came into school with this vision of working for a clean-tech
start-up," he told me. "But I also had this idea floating around in the
back of my head that we're moving too slowly to actually stop emissions in time. So what do we do with all the carbon that's in the air?"
He started talking to scientists and policy experts at Berkeley. What he
learned shocked him.

"People told me, 'The models show this major need for negative
emissions,'" he recalled. "'But we don't really know how to do that,
nor is anyone really thinking about it.' I was someone who'd been in
the business and policy world, and I was, like, wait a minute—what?"

Business school taught Deich to think in terms of case studies. One
that seemed to him relevant was solar power. Photovoltaic cells have
been around since the nineteen-fifties, but for decades they were prohibitively expensive. Then the price started to drop, which increased
demand, which led to further price drops, to the point where today, in
many parts of the world, the cost of solar power is competitive with
the cost of power from new coal plants.

"And the reason that it's now competitive is that governments decided to do lots and lots of research," Deich said. "And some countries,
like Germany, decided to pay a lot for solar, to create a first market.
And China paid a lot to manufacture the stuff, and states in the U.S.
said, 'You must consume renewable energy,' and then consumers said,
'Hey, how can I buy renewable energy?'"

As far as he could see, none of this—neither the research nor the
creation of first markets nor the spurring of consumer demand—was
being done for carbon removal, so he decided to try to change that.
Together with a Berkeley undergraduate, Giana Amador, he founded
the center in 2015, with a hundred-and-fifty-thousand-dollar grant
from the university. It now has an annual budget of about a million

dollars, raised from private donors and foundations, and a staff of seven. Deich described it as a "think-and-do tank."

"We're trying to figure out: How do we actually get this on the agenda?" he said.

A compelling reason for putting carbon removal on "the agenda" is that we are already counting on it. Negative emissions are built into the IPCC scenarios and the climate agreements that rest on them.

But everyone I spoke with, including the most fervent advocates for carbon removal, stressed the huge challenges of the work, some of them technological, others political and economic. Done on a scale significant enough to make a difference, direct air capture of the sort pursued by Carbon Engineering, in British Columbia, would require an enormous infrastructure, as well as huge supplies of power. (Because CO_2 is more dilute in the air than it is in the exhaust of a power plant, direct air capture demands even more energy than CCS.) The power would have to be generated emissions-free, or the whole enterprise wouldn't make much sense.

"You might say it's against my self-interest to say it, but I think that, in the near term, talking about carbon removal is silly," David Keith, the founder of Carbon Engineering, who teaches energy and public policy at Harvard, told me. "Because it almost certainly is cheaper to cut emissions now than to do large-scale carbon removal."

BECCS doesn't make big energy demands; instead, it requires vast tracts of arable land. Much of this land would, presumably, have to be diverted from food production, and at a time when the global population—and therefore global food demand—is projected to be growing. (It's estimated that to do BECCS on the scale envisioned by some below-two-degrees scenarios would require an area larger than India.) Two researchers in Britain, Naomi Vaughan and Clair Gough, who recently conducted a workshop on BECCS, concluded that "assumptions regarding the extent of bioenergy deployment that is possible" are generally "unrealistic."

For these reasons, many experts argue that even talking (or writing articles) about negative emissions is dangerous. Such talk fosters the impression that it's possible to put off action and still avoid a crisis, when it is far more likely that continued inaction will just produce a larger crisis. In "The Trouble with Negative Emissions," an essay that ran last year in *Science,* Kevin Anderson, of the Tyndall Centre for Climate Change Research, in England, and Glen Peters, of the climate research center in Oslo, described negative-emissions technologies as a "high-stakes gamble" and relying on them as a "moral hazard par excellence."

We should, they wrote, "proceed on the premise that they will not work at scale."

Others counter that the moment for fretting about the hazards of negative emissions—moral or otherwise—has passed.

"The punch line is, it doesn't matter," Julio Friedmann, the former principal deputy assistant energy secretary, told me. "We actually need to do direct air capture, so we need to create technologies that do that. Whether it's smart or not, whether it's optimized or not, whether it's the lowest-cost pathway or not, we know we need to do it."

"If you tell me that we don't know whether our stuff will work, I will admit that is true," Klaus Lackner said. "But I also would argue that nobody else has a good option."

One of the peculiarities of climate discussions is that the strongest argument for any given strategy is usually based on the hopelessness of the alternatives: this approach *must* work, because clearly the others aren't going to. This sort of reasoning rests on a fragile premise—what might be called solution bias. There has to be an answer out there somewhere, since the contrary is too horrible to contemplate.

Early last month, the Trump administration announced its intention to repeal the Clean Power Plan, a set of rules aimed at cutting power plants' emissions. The plan, which had been approved by the Obama administration, was eminently achievable. Still, according to the current administration, the cuts were too onerous. The repeal of the plan is likely to result in hundreds of millions of tons of additional emissions.

A few weeks later, the United Nations Environment Programme released its annual Emissions Gap Report. The report labeled the difference between the emissions reductions needed to avoid dangerous climate change and those that countries have pledged to achieve as "alarmingly high." For the first time, this year's report contains a chapter on negative emissions. "In order to achieve the goals of the Paris Agreement," it notes, "carbon dioxide removal is likely a necessary step."

As a technology of last resort, carbon removal is, almost by its nature, paradoxical. It has become vital without necessarily being viable. It may be impossible to manage and it may also be impossible to manage without.

Originally published in *The New Yorker,*
November 20, 2017.

Update: In 2023, Carbon Engineering was purchased by Occidental Petroleum for $1.1 billion. Occidental is building a plant in Texas that, when finished, is supposed to suck five hundred thousand tons of carbon dioxide out of the air each year.

RECALL OF THE WILD

Can People Re-create Nature?

FLEVOLAND, WHICH SITS more or less in the center of the Netherlands, half an hour from Amsterdam, is the country's newest province, a status that is partly administrative and partly existential. For most of the past several millennia, Flevoland lay at the bottom of an inlet of the North Sea. In the nineteen-thirties, a massive network of dams transformed the inlet into a freshwater lake, and in the nineteen-fifties a drainage project, which was very nearly as massive, allowed Flevoland to emerge out of the muck of the former seafloor. The province's coat of arms, drawn up when it was incorporated, in the nineteen-eighties, features a beast that has the head of a lion and the tail of a mermaid.

Flevoland has some of Europe's richest farmland; its long, narrow fields are planted with potatoes and sugar beets and barley. On each side of the province is a city that has been built from scratch: Almere in the west and Lelystad in the east. In between lies a wilderness that was also constructed, Genesis-like, from the mud.

Known as the Oostvaardersplassen, a name that is pretty much un-pronounceable for English speakers, the reserve occupies fifteen thousand almost perfectly flat acres on the shore of the inlet-turned-lake. This area was originally designated for industry; however, while it was still in the process of drying out, a handful of biologists convinced the Dutch government that they had a better idea. The newest land in Europe could be used to create a Paleolithic landscape. The biologists set about stocking the Oostvaardersplassen with the sorts of animals that would have inhabited the region in prehistoric times—had it not at that point been underwater. In many cases, the animals had been exterminated, so they had to settle for the next best thing. For example, in place of the aurochs, a large and now extinct bovine, they brought in Heck cattle, a variety specially bred by Nazi scientists. (More on the Nazis later.) The cattle grazed and multiplied. So did the red deer, which were trucked in from Scotland, and the horses, which were imported from Poland, and the foxes and the geese and the egrets. In fact, the large mammals reproduced so prolifically that they formed what could, with a certain amount of squinting, be said to resemble the great migratory herds of Africa; the German magazine *Der Spiegel* has called the Oostvaardersplassen "the Serengeti behind the dikes." Visitors now pay up to forty-five dollars each to take safari-like tours of the park. These are especially popular in the fall, during rutting season.

Such is the success of the Dutch experiment—whatever, exactly, it is—that it has inspired a new movement. Dubbed Rewilding Europe, the movement takes the old notion of wilderness and turns it inside out. Perhaps it's true that genuine wildernesses can only be destroyed, but new "wilderness," what the Dutch call "new nature," can be created. Every year, tens of thousands of acres of economically marginal farmland in Europe are taken out of production. Why not use this land to produce "new nature" to replace what's been lost? The same basic idea could, of course, be applied outside of Europe—it's been proposed, for example, that depopulated expanses of the American Midwest are also candidates for rewilding.

I visited the Oostvaardersplassen during a stretch of very blue days in early fall. As it happened, two film crews, one Dutch and the other French, were also there. The French crew, whose credits include the international hit *Winged Migration,* was scouting the reserve for possible use in an upcoming feature about the history of Europe as seen through the eyes of other species. The Dutch crew was finishing up a full-length nature documentary. One afternoon, we all got into vans and drove to the middle of the park. A stiff breeze was blowing, as it almost always does near the North Sea. We passed a marshy area covered in reeds, which nodded in the wind. Ducks bobbed in a pond. Farther on, where the land grew drier, the reeds gave way to grass. We passed a herd of red deer and some aurochs wannabes, and the carcass of a deer, which had been picked almost clean by foxes and ravens. (The Dutch crew had filmed the scavenging with a time-lapse camera.) Eventually, we came to a herd of about a thousand wild—or, at least, feral—horses. They whinnied and cantered and shook their heads. The horses were an almost uniform buff color, and the breeze lifted their manes, which were dark brown. We all piled out of the vans. The horses seemed not to notice us, though we were just a few yards away.

"*Ah, c'est joli ça!*" the French exclaimed. A flock of black-and-white barnacle geese rose into the air and then, a moment later, a yellow train clicked by, carrying passengers from Almere to Lelystad or, perhaps, vice versa. A few members of the French crew had brought along video cameras. As they panned across the horses—at the edge of the herd, a mare nuzzled a foal that couldn't have been more than two or three days old—I wondered what they would do with the high-voltage power lines in the background. It occurred to me that, like so many postmodern projects, the Oostvaardersplassen was faintly ridiculous. It was also, I had to admit, inspiring.

IF ONE PERSON could be said to be responsible for the Oostvaard-ersplassen, it is an ecologist named Frans Vera. Vera, who is sixty-three, has gray hair, a gray beard, and a cheerfully combative manner. He spent most of his adult life working for one or another branch of the Dutch government and now works for a private foundation, of which, as far as I could tell, he is the sole employee. Vera picked me up one day at my hotel in Lelystad, and we drove over to the reserve's administrative offices, where we had a cup of coffee in a room decorated with the mounted head of a very large black Heck bull.

Vera explained that he first became interested in the Oostvaarders-plassen in the late nineteen-seventies. At that point, he had just graduated from university, in Amsterdam, and was unemployed. He read an article about some graylag geese that had appeared in the reclaimed area, which was then a boggy no-man's-land. The geese kept the vegetation low by chomping on it and in this way maintained their marshy habitat. Vera was an avid bird-watcher, and the story intrigued him. He wrote his own article, arguing that the place ought to be turned into a nature preserve. Soon afterward, he got a job with the Dutch forestry agency.

In the late seventies, the prevailing view in the Netherlands was— and, to a certain extent, it still is—that nature was something to be managed, like a farm. According to this view, a preserve needed to be planted, pruned, and mowed, and the bigger the preserve, the more intervention was required. Vera chafed at this notion. The problem, he decided, was that Europe's large grazers had been hunted to oblivion. If they could be restored, then nature could take care of itself. This theory, coming from a very junior civil servant, was not particularly popular.

"Mostly there's no trouble as long as you are within the borders of an accepted paradigm," Vera told me. "But be aware when you start to discuss the paradigm. Then it starts to be only 25 percent discussion of facts and 75 percent psychology. The thing I most often heard was, 'Who do you think you are?'" Undaunted, Vera kept pushing. He had a few allies at various government ministries, and one of them arranged for him to get the money to buy some Heck cattle. In 1983, while the future of the Oostvaardersplassen was still being debated, Vera acquired the cows from Germany, although he had not yet secured permission from the governing authorities to release them.

"I bought them and I was standing here with the trucks," he recalled happily. "And they were so angry!" This first group of Heck cattle was not allowed onto the site, but a second group, acquired some months later, was let in. The following year, Vera bought forty Konik horses from Poland. Koniks are believed to be descended from tarpans, one of the world's last subspecies of truly wild horse, which survived in eastern Europe into the nineteenth century. (Practically all the horses that are called "wild" today are, in fact, the offspring of domesticated horses that were, at some point or another, let loose.) Red deer, which are closely related to what Americans call elk, were brought in during the nineteen-nineties.

Meanwhile, other animals were finding their way to the Oostvaard-ersplassen on their own. Foxes arrived, as did muskrats, which in Europe count as an invasive species. Buzzards and goshawks and gray herons and kingfishers and kestrels turned up. A pair of very large white-tailed eagles swooped in and built their nest in an improbably small tree. In 2005, a rare black vulture appeared, but after a few months in residence it wandered onto the railroad tracks, where it was hit by a train. (The rail line runs along the southern edge of the preserve.) Vera's dream is that one day the Oostvaardersplassen will be connected to other nature reserves in the Netherlands—a plan that has been partly but never fully funded—and that this will, in turn, allow it to attract wolves. Wolves were extirpated from most of western Europe more than a century ago, but, owing to stringent protections put in place over the past few decades, they have recently been making a comeback in countries like Germany and France. (Two packs, with about ten wolves each, now live within forty miles of Berlin.) Last year, a wolf believed to be the first seen in Holland since the eighteen-sixties was spotted about seventy miles southeast of the Oostvaardersplassen, in the town of Duiven.

"That is probably unimaginable for people in the United States—having wolves in the Netherlands," Vera said. "But it is the future."

AFTER WE HAD finished our coffee, we got into a truck and drove through the gates of the preserve. So effectively have the cows and the horses and the deer kept the place grazed that there was barely a bush to be seen—just acre after very flat acre of clipped grass, like a bowling green. We passed a few groups of deer and a fox that looked back at us with pale, glittering eyes. Vera stopped the truck at a lookout built on stilts. We climbed up a narrow ladder. "This is a window that shows us how the Netherlands looked thousands of years ago," he said, gesturing at the grassland below.

A corollary of Vera's theory about large grazers is a second hypothesis, which he has pushed even more vigorously than the first, if that's possible. Among ecologists, the prevailing view of Europe in its natural, which is to say preagrarian, state is that it was heavily forested. (The continent's last stands of old-growth forest are found on the border of Poland and Belarus, in the Białowieża Forest, which the author Alan Weisman has described as a "relic of what once stretched east to Siberia and west to Ireland.")

Vera argues that, even before Europeans figured out how to farm, the continent was more of a parklike landscape, with large expanses of open meadow. It was kept this way, he maintains, by large herds of herbivores—aurochs, red deer, tarpans, and European bison. (The bison, also known as wisents, were hunted nearly to extinction by the late eighteen-hundreds.) Vera has written up his argument in a dense, five-hundred-page treatise that has received a good deal of attention from European naturalists, not all of it favorable. A botany professor at Dublin's Trinity College, Fraser Mitchell, has written that an analysis of ancient pollen "forces the rejection of Vera's hypothesis." Vera, for his part, rejects the rejection, arguing that, precisely because they ate so much grass, the aurochs and the wisents skewed the pollen record. "That is a scientific debate that is still going on," he told me.

Like the rest of Flevoland, the Oostvaardersplassen lies about fifteen feet below sea level and is protected from flooding by a series of thick earthen dikes. As a result, when you are standing in the park, the lake, known as the Markermeer, is above you, which produces the vertiginous sense of a world upside down. In the lovely weather, the Markermeer was filled with sailboats; these seemed to be hovering above the horizon, like zeppelins.

"What we see here is that, instead of what many nature conservationists think—that something that is lost is lost forever—you can have the conditions to have it redeveloped," Vera told me. "So this is the ultimate proof. There's no bird here who says, 'I won't breed here, because it's unnatural—it's four and a half meters below sea level, and I never did that.'" We drove on, and stopped to take a look at the nest built by the white-tailed eagles, another animal that only very narrowly avoided extinction. The eagles showed up in the Oostvaarders-plassen in 2006 and became the first pair to breed in the Netherlands since the Middle Ages. Their nest—empty at the time of my visit—was an extraordinary structure, made out of sticks and nearly the size of an armchair. It seemed ready to topple the scrawny tree it was perched in. Vera was particularly pleased with the eagles, because several ornithologists had told him the birds would nest only in very tall, mature trees, of which the Oostvaardersplassen has none.

"Many so-called specialists thought this would be impossible," he said. "The eagles had a different opinion."

Access to the Oostvaardersplassen by humans is strictly controlled, and that morning neither of the film crews was there and no tours were out, so Vera and the animals and I pretty much had the place to ourselves. The quiet was interrupted only by the squawking of the geese and the clatter of an occasional train. We continued west, skirting a herd of red deer. A dead horse was lying in the middle of the herd. Its chest was bloated, and there was a large dark hole where its anus once had been. Vera speculated that it had been made by foxes trying to get at the horse's entrails.

Like genuinely wild animals, those in the Oostvaardersplassen are expected to fend for themselves. They are not fed or bred or vaccinated. Also like wild animals, they often die for lack of resources; for the large herbivores in the reserve, the mortality rate can approach 40 percent a year. From a public relations point of view, this is far and away the most controversial aspect of Vera's scheme. When the weather is harsh, there's widespread starvation in the preserve, which provides gruesome images for Dutch TV. Often the dying animals are shown huddled up against the fences of the Oostvaardersplassen, a scene that invariably leads to comparisons with the Holocaust.

"You can't have a discussion without the Second World War coming up," Vera told me. "It's really sick-making." In the fall of 2005, the controversy became so heated that the Dutch government appointed a committee—the International Committee on the Management of Large Herbivores in the Oostvaardersplassen, or ICMO—to look into the matter. ICMO recommended a policy of "reactive culling," under which the animals would be monitored over the winter, and those that seemed too weak to survive until spring would be shot.

Michael Coughenour, a research scientist at the Natural Resource Ecology Laboratory at Colorado State University, was a member of ICMO. He told me that while it was difficult to compare mortality rates at the Oostvaardersplassen to those in a place like the Serengeti, "severe-winter die-offs are a natural thing."

"I didn't see anything that looked bad to me," he went on, referring to a visit the committee members made to the Oostvaardersplassen. "I think it's a great experiment to let it run and see what happens."

Even though ICMO's recommendations were adopted, many critics were not satisfied, and in 2006 a Dutch animal welfare association sued the managers of the Oostvaardersplassen for what it alleged was continuing mistreatment. The group lost the case, appealed, and lost again. Then, in the winter of 2010, an unusually cold one in northern Europe, a Dutch news program aired a segment on the Oostvaardersplassen that showed an emaciated deer stumbling into a half-frozen pond and drowning. A public outcry ensued, prompting an "emergency" debate in Parliament.

"It's an illusion to think we can go back to primordial times, dressed in bear furs and floating around in hollowed-out trees," the MP who led the debate, Henk Jan Ormel, said. "The world of today looks very different, and we shouldn't make the animals of the Oostvaardersplassen bear the burden of this."

"It became political," Sip van Wieren, a professor of ecology at Wageningen University, told me. "*Very* political." A second ICMO was convened. This one recommended a policy of "early reactive culling," under which the animals that were deemed unlikely to survive the winter would be shot in the fall. How exactly the rangers at the Oostvaardersplassen were supposed to figure out in November which animals would be starving by February was left rather vague.

When I visited, in September, the number of grazers in the park was at its annual peak, with more than three thousand deer, a thousand horses, and three hundred Heck cattle. Eventually, it is hoped, birth rates in the Oostvaardersplassen will decline, and the population will reach some kind of equilibrium, but in the meantime the shooting continues. Vera and I came upon a group of cows sunning themselves near a dead tree. They regarded us warily, through glassy black eyes. The adults looked fearfully robust, but some of the calves seemed a bit shaky; within a few months, I figured, they'd probably be carcasses. Vera told me that he viewed "early reactive culling" as an arrangement whose only real beneficiaries were humans; as far as the ungulates were concerned, he thought, starving to death was a very peaceful way to go.

"It only has to do with the acceptance of people," he said, "and nothing, in my mind, to do with the suffering of animals."

THERE ARE MORE than 1.5 billion cows in the world today, and all of them are believed to be descended from the aurochs—*Bos primigenius*—which once ranged across Europe, much of Asia, and parts of the Middle East. Aurochs were considerably more impressive beasts than domesticated cattle. Julius Caesar described them as being just "a little below the elephant in size," with "strength and speed" that was "extraordinary." (It is unlikely that he ever actually saw one.) More recent estimates suggest that males were nearly six feet high at the withers and females five feet. By Roman times, humans had so diminished the aurochs population numbers that the animals were missing from most of their former habitat.

By the fifteen hundreds, the only place they could still be found in the wild was in the Polish Royal Forests, west of Warsaw. The animals there were understood to be extremely rare, and special gamekeepers were hired to protect them. But their numbers continued to dwindle. In 1557, some fifty aurochs were counted. Forty years later, only half that many remained, and by 1620 only one aurochs—a female—was left. She died in 1627. The aurochs thus earned, as the Dutch writer Cis Van Vuure has put it, "the dubious honor of being the first documented case of extinction." (The next case was the dodo, four decades later.)

The aurochs was essentially forgotten until the early twentieth century, when a spate of scientific papers on the animal appeared. In the nineteen-twenties, two German brothers, Heinz and Lutz Heck, both zoo directors, decided to try to breed back the aurochs, using the genetic material that had been preserved in domesticated cattle. This was, of course, long before DNA testing—or even the discovery of DNA. To guide their efforts, the brothers mainly relied on old pictures of aurochs, many of them drawn by people with no firsthand knowledge of the animal. The brothers chose different kinds of cows for their breeding efforts: Heinz, who directed the zoo in Munich, crossed, among other breeds, Scottish Highland cattle and German Anglers, while Lutz, the director of the Berlin Zoo, mixed Spanish fighting cattle with Corsican and Camargue cattle. Nevertheless, the two claimed that their efforts had produced similar results, which, they argued, proved that "the fundamental principle of breeding back was correct." Even though he continued to crossbreed his crossbreeds, Heinz decided that the project had been successfully completed. "The wild bull, the aurochs, lives again," he wrote.

Not long afterward, the project became tangled up in German politics. In 1938, Lutz, a committed Nazi, was appointed to the Third Reich's Forest Authority. His idea of breeding back the aurochs dovetailed neatly with the Nazis' scheme of restoring Europe, through selective human breeding, to its mythic Aryan past. Lutz sent some of his "aurochs" to the Rominten Heath, in East Prussia—now Poland—where Hermann Göring had his favorite hunting lodge. Other Heck-bred cows were installed on the grounds of Göring's estate north of Berlin. Most—perhaps all—of these animals were killed toward the end of the Second World War. (According to Clemens Driessen, a Dutch academic who has studied the Heck brothers, Göring personally shot some of the cattle on his estate as the Soviets bore down on Berlin.) But some Heck cattle at the Munich zoo and in parks in Augsburg, Münster, and Duisburg survived.

Over the years, even as Heck cattle have been raised, uneventfully, in once-Nazi-occupied nations like the Netherlands—it's the descendants of the Munich-bred cows that now graze the Oostvaardersplassen—they've never managed to shake their Fascist associations. Many regard them as a sort of veterinary version of the Hitler Diaries—half horror, half joke. Not long ago, when a British farmer imported some Heck cattle from Belgium, the story made national news.

"NAZI 'SUPER-COWS' SHIPPED TO DEVON FARM," *The Guardian* reported.

"THE HERD REICH," ran the headline in *The Sun*.

As more aurochs remains have been unearthed and more sophisticated research has been done on them, it's become clear that the Heck brothers' creation is a far cry from the original—Heck cattle are too small, their horns have the wrong shape, and the proportions of their bodies are off. All of this has led to a new, de-Nazified effort to back-breed the aurochs. This project is based in the Dutch city of Nijmegen, about fifty miles southeast of Amsterdam, and is entirely independent of the Oostvaardersplassen. Still, it reflects much the same can-do, "what is lost is not lost forever" approach to conservation. So while I was in the Netherlands I decided to go for a visit.

"WATCH OUT," Henri Kerkdijk warned. It was another surprisingly blue day, and we were tromping through a weedy field toward a line of trees. I looked back at him, which turned out to be a mistake, because at that moment I stepped into a large pile of cow shit. As I scraped it from my shoes, I wondered how much bigger the pile would have been had it been produced by an actual aurochs.

Standing in the shade of the trees were about a dozen cows of varying color and size. Kerkdijk pointed to two black bulls bent over a patch of grass. The first was called Manolo Uno. He was two years old and not yet fully grown, but already he measured almost five feet at the withers. He had a grayish muzzle, a light stripe down his back, and forward-tilting horns that reminded me of Ferdinand's. I have no idea how closely he resembled an actual aurochs; certainly, though, he seemed a very imposing beast, larger and more menacing-looking than the Heck cattle at the Oostvaardersplassen. The second bull, Rocky, was a year younger than Manolo but almost as big. This Kerkdijk took as a particularly promising sign. "That one's going to be really tall," he said.

Four years ago, Kerkdijk teamed up with an environmental consultant named Ronald Goderie to start the TaurOs program, the stated goal of which is to give "the rebuilding of the aurochs a serious try." (In a recent write-up of the effort, the two men dismiss Heck cattle as "considered by experts to be a failure.") At the point that I met with them, the project had generated nearly a hundred calves, of which Manolo Uno and Rocky had been deemed the most aurochs-like. To create the calves, Kerkdijk and Goderie had crossed several so-called primitive cattle breeds—varieties developed hundreds, even thousands, of years ago, and therefore more likely to retain aurochs-like features. Manolo, for example, represents a cross between an Italian breed known as Maremmana primitivo and a Spanish breed known as Pajuna. At two, he was old enough to be crossbred himself. But he had refused to part with any of his semen for the purpose of artificial insemination, a demurral that Kerkdijk took as evidence of his virility and a further positive sign.

Ninety years after the Heck brothers' attempt, the basic idea behind back-breeding remains pretty much the same. If different breeds of primitive cattle preserve different stretches of the aurochs's genetic material, then reassembling those stretches should produce something close to—though not exactly like—the original. (Kerkdijk and Goderie have decided that their new animal should be called not an aurochs but a "tauros.") Scientists in England and Ireland have succeeded in sequencing a small subset of the aurochs's DNA—its mitochondrial DNA—using a seven-thousand-year-old bone that was found in a cave in Derbyshire. Other scientists have been approached about sequencing the entire genome. When—or, really, if—this work is completed, it should be possible to gauge how close a calf comes to an authentic aurochs by analyzing a blood sample or a bit of saliva.

According to the timetable Kerkdijk and Goderie have drawn up, herds of "tauroses" should be ready by around 2025. By that point, the two expect that large tracts of Europe will have been rewilded, and the animals will be allowed to roam across them. How the intervening years' worth of breeding and cross-breeding and genetic evaluation will be funded remains a bit murky. Currently, the project is supported in part by renting cows to nature parks and in part by butchering them. The meat is marketed as "wild beef," and it commands a premium in Amsterdam, where it is available only to customers who sign up for delivery in advance. Kerkdijk said that "wild beef" sales had risen dramatically over the last year or so, owing to interest in the tauros. I asked if I could try some.

"Did you bring your bow and arrow?" Goderie asked.

LIKE SO MUCH in Europe today, the term *rewilding* is an American import. It was coined in the nineteen-nineties, and first proposed as a conservation strategy by two biologists, Michael Soulé, now a professor emeritus at the University of California at Santa Cruz, and Reed Noss, a research professor at the University of Central Florida. According to Soulé and Noss, the problem with most conservation plans was that they aimed to protect what exists. Yet what exists is often just a shadow of what once was. In most of the United States, large predators like wolves and cougars have been wiped out. Without top predators, the two argued, ecosystems no longer really function as systems.

"A cynic might describe rewilding as an atavistic obsession," they wrote. "A more sympathetic critic might label it romantic. We contend, however, that rewilding is simply scientific realism." According to Soulé and Noss, rewilding demanded, in addition to predators, the establishment of large, strictly protected "core" reserves, and migratory corridors linking one to the next. They summarized their formula as "the three C's: cores, corridors, and carnivores." These ideas are now considered mainstream by conservation biologists, even those who would not necessarily describe themselves as proponents of rewilding.

In 2005, a dozen biologists took the concept of rewilding one step further. In an article published in the journal *Nature,* the group presented a plan for what it called "Pleistocene rewilding."

When humans arrived in North America, some thirteen thousand years ago, toward the end of the last ice age, they killed off most of the continent's largest mammals, leaving key ecological roles unfilled. The Pleistocene rewilders proposed finding substitute animals that could serve in their place. For instance, African or Asian elephants could be let loose to make up for the long-lost woolly mammoth. Similarly, Bactrian camels, which are native to the steppes of Central Asia, could take up the slack left by the vanished North American *Camelops.* The authors—almost all of them were academics—envisioned a series of small-scale experiments leading up to the creation of "one or more 'ecological history parks,'" which would cover "vast areas of economically depressed parts of the Great Plains." In these huge "history parks," elephants, camels, and African cheetahs—to replace the missing American cheetah—would roam freely. The ecologists called their plan "an optimistic alternative" to what was otherwise likely to be a future filled with "ever more pest-and-weed dominated landscapes" and "the extinction of most, if not all, large vertebrates."

The lead author of the *Nature* article, Josh Donlan, now runs a nonprofit group called Advanced Conservation Strategies and is a visiting fellow at Cornell. He characterized reactions to Pleistocene rewilding as "bimodal."

"People either loved it or hated it, both in the scientific community and in the public," he told me. In the United States, Pleistocene rewilding never got very far; the only practical step that's been taken has been the reintroduction to private land in New Mexico of a giant tortoise known as the Bolton tortoise. (The Bolton tortoise, which disappeared from what's now the U.S. about eight thousand years ago, survived south of the border in very small numbers.) As it happened, though, a Russian scientist named Sergey Zimov had a similar idea. Also in 2005, he published an article in *Science* describing an experimental preserve in Siberia that he had set up and named the Pleistocene Park. Zimov's aim was to show that the area, which ten thousand years or so ago supported great herds of large mammals, was still capable of doing so.

"We are not trying exactly to reconstruct the mammoth steppe ecosystem, because we don't have the mammoth," Zimov told me recently by phone from St. Petersburg. "But we are trying to reconstruct the highly productive steppe ecosystem." Zimov brought in reindeer and a breed of very cold-hardy horses known as Yakutians. A few years ago, he imported five European bison to the park, but only one—a male—survived the second winter. "Now we are looking for girlfriends," Zimov said. Several musk oxen were also brought in, but they, too, were all males. "We also search females for them," Zimov told me. The Pleistocene Park, which is in northeastern Siberia, is so remote that almost no one who isn't conducting research there has ever visited it.

As Europeans have taken up the term, *rewilding* has shifted its meaning yet again. The concept has become at once less threatening and more gastronomically appealing: it is expected that visitors to the continent's rewilded regions will be able to enjoy not just the safari-like tours but also the local cuisine. (One park in Portugal in the process of "rewilding" offers its own brand of olive oil.)

Rewilding Europe, the group that is pushing the concept most vigorously, was founded three years ago by two Dutchmen, a Swede, and a Scot. One of the Dutchmen, Wouter Helmer, lives not far from the field where Manolo and Rocky are pastured, and the day after I visited the bulls I went to meet him at his house, which is at the edge of a park, in a small clearing that made me think of Goldilocks.

Helmer explained that the goal of Rewilding Europe was, in effect, to create giant versions of the Oostvaardersplassen, each at least fifteen times as large. "Frans Vera always says, 'If the Dutch can do it, everyone can do it,'" he told me. To get the project started, the group has raised more than six million euros—roughly seven and a half million dollars—much of it from the Dutch post-code lottery, which might be compared to the New York State lottery, except that the proceeds go to charity. Last year, after receiving twenty applications from organizations across the continent, the group chose five regions to serve as what it calls "model rewilding areas"—a part of the Danube Delta, spanning the border of Romania and Ukraine; an area in the southern Carpathian Mountains, also known as the Transylvanian Alps; and areas in the eastern Carpathians, the mountains of Croatia, and the western Iberian Peninsula. One quality these areas share is that fewer and fewer people want to live in them.

"There's no economy in big parts of Europe," Helmer told me. "We think it's a window of opportunity." The idea is to rewild the areas by connecting existing reserves with tracts of abandoned land and working farms whose owners can be persuaded to let a herd of aurochs (or tauroses) wander across their property. (The lure for landowners is supposed to be an influx of tourists, who will come and open their wallets.)

Helmer stressed to me that Rewilding Europe was not particularly concerned about whether the new landscape that would be created would resemble the ancient one that had been altered or destroyed. "We're not looking backward but forward," he said at one point.

"We try to avoid too much discussion of wilderness," he observed at another. "For us, that is not the most important thing—at the end will this be a wilderness or not? It will be wilder than it was, and that's what matters."

ONE MORNING not long after this, I found myself sitting in a small hut, staring at a pile of dead chickens. The chickens had pure white feathers that were matted with blood, and they lay with their half-severed heads and rigid legs tilted at grotesque angles. After a while, a half-dozen griffon vultures settled into a nearby tree. Griffon vultures are large birds with light-colored faces and dark bodies, and the group in the tree resembled a gathering of harpies. A little while later, a pair of black vultures showed up and began circling overhead. Black vultures are even larger than griffons, with wingspans that can reach ten feet. They are majestic, funereal-looking birds, and watching them feels like a premonition of one's own death. The chickens had been laid out as part of a supplementary feeding program for the birds, who, it seemed, were not hungry. The black vultures continued to circle, the griffon vultures continued to sit in the tree, and the small hut grew stuffier. After a few hours, my companion, Diego Benito, decided that the spectacle we had come to see was not going to take place, and so, disappointed, we left.

Benito runs a thirteen-hundred-acre nature preserve in far western Spain called the Campanarios de Azaba. The preserve is part of the Rewilding Europe "model area" in western Iberia, and of the five areas it's the easiest to get to. Nevertheless, the trip there involves a four-hour drive from Madrid, through the provinces of Ávila and Salamanca.

Since the vultures weren't cooperating, Benito suggested we tour the rest of the reserve. Until fairly recently, the place had been a farm, and it was dotted with oak trees whose acorns had gone to fattening pigs. It was hot and dry as we crunched along through the underbrush. Even though I knew the nearest town wasn't more than a few miles away, the terrain seemed empty enough to get lost in, and I was reminded of a time in the New Mexico desert when I'd read a trail map wrong and found myself walking in circles. We encountered some very handsome horses, which, Benito told me, belonged to a rare and ancient Spanish breed known as Retuertas. Farther on, we came to a fenced-in area filled with a network of small but clearly man-made tunnels. These, Benito explained, had been dug for the benefit of rabbits, which in Spain—and, indeed, throughout Europe—have been decimated by a disease known as myxomatosis. The myxoma virus was purposefully introduced on a private estate in France as a rabbit control measure in the nineteen-fifties and has since spread across the continent. (The loss of rabbits has led to a decline in animals that prey on rabbits, like the Iberian lynx, which is now considered to be critically endangered.) The fences were supposed to protect some reintroduced rabbits from foxes, but the rabbits had refused to stay put, so now the enclosures were empty. The same was true of a series of circular platforms that had been erected in some oak trees as nesting sites for black storks. The black storks hadn't been interested in them.

"You can't be a hundred percent sure of success, because wild animals are wild animals," Benito told me. We went looking for some Sayaguesa cows that had recently been purchased with Rewilding Europe money, but they seemed to be avoiding us. Sayaguesas are another primitive breed of interest to the TaurOs program, an enterprise that Benito told me he was eager to get involved in. "If you want to sell a product, you have to have a story," he said.

That afternoon, after a lunch of local (and quite tasty) pork cutlets, we drove out of the reserve to the top of a nearby mountain. Along the way, we passed through a couple of villages that, Benito explained, were in the process of disappearing; the schools had closed for lack of children and only the old people remained. In one of the towns, La Encina, we stopped to meet the mayor, a slight, elderly man named José Maria. According to Maria, the number of residents in La Encina had dropped by more than 50 percent in just the past fifteen years. He was enthusiastic about the idea of rewilding, he said, because it had "a lot of potential to bring tourists." From the top of the mountain, we could see across to Portugal, some fifteen miles away. The valley was a patchwork of brown fields, pine forests that had been planted during the Franco era, and evenly spaced oaks of the sort I'd seen at the preserve. According to a brochure that Wouter Helmer had given me, the entire region was ripe for rewilding, owing to "rural depopulation"; the aim was to transform at least a thousand square kilometers, or two hundred and fifty thousand acres. I tried to imagine the whole valley converted into an Iberian version of the Oostvaardersplassen. Certainly it was a lot less populated than the outskirts of Amsterdam. Still, I realized, I wasn't sure what I was supposed to be envisioning. The pine plantations could never be considered wild: Would they have to go? What about the pruned oaks, and the pigs that were still snuffling around them for acorns, and the brown fields, and all the tiny, dying towns waiting for an influx of tourists?

One of the appeals of rewilding is that it represents a proactive agenda—as Josh Donlan and his Pleistocene rewilding colleagues put it, a hopeful alternative to just sitting around, mourning what's been lost. In a rewilded world, even extinction need not be considered irrevocable; the aurochs will lie down with the lynx, and the deer and the elephants will roam. On a planet increasingly dominated by people—even the deep oceans today are being altered by humans—it probably makes sense to think about wilderness, too, as a human creation. The more I saw, the more I understood why Europeans, in particular, were attracted to the idea, and the more I wanted to be convinced that it could work. But, as I looked back toward the Campanarios de Azaba, I thought of the vacant rabbit tunnels and the empty platforms built for the storks, and I wasn't at all sure.

It was dusk by the time we headed down the mountain. Benito got a call on his cell phone from a local farmer who had a dead pig he thought the vultures might be interested in. On our way back, we stopped by to see what had happened to the chickens. Every one of them was gone, including the bones.

Originally published in *The New Yorker,*
December 24 and 31, 2012.

Update: The aurochs genome was sequenced in 2015. In 2024, some "tauroses" were released in Portugal.

PART FOUR

ALL WE CAN SAVE

THE CATASTROPHIST

Profile of James Hansen,
the "Father of Global Warming"

A FEW MONTHS AGO, James Hansen, the director of NASA's Goddard Institute for Space Studies, in Manhattan, took a day off from work to join a protest in Washington, D.C. The immediate target of the protest was the Capitol Power Plant, which supplies steam and chilled water to congressional offices, but more generally its object was coal, which is the world's leading source of greenhouse gas emissions. As it happened, on the day of the protest it snowed. Hansen was wearing a trench coat and a wide-brimmed canvas boater. He had forgotten to bring gloves. His sister, who lives in D.C. and had come along to watch over him, told him that he looked like Indiana Jones.

The march to the power plant was to begin on Capitol Hill, at the Spirit of Justice Park. By the time Hansen arrived, thousands of protesters were already milling around, wearing green hard hats and carrying posters with messages like "Power Past Coal" and "Clean Coal Is Like Dry Water." Hansen was immediately surrounded by TV cameras.

"You are one of the preeminent climatologists in the world," one television reporter said. "How does this square with your science?"

"I'm trying to make clear what the connection is between the science and the policy," Hansen responded. "Somebody has to do it."

The reporter wasn't satisfied. "Civil disobedience?" he asked, in a tone of mock incredulity. Hansen said that he couldn't let young people put themselves on the line, "and then I stand back behind them."

The reporter still hadn't got what he wanted: "We've heard that you all are planning, even hoping, to get arrested today. Is that true?"

"I wouldn't hope," Hansen said. "But I do want to draw attention to the issue, whatever is necessary to do that."

Hansen, who is sixty-eight, has greenish eyes, sparse brown hair, and the distracted manner of a man who's just lost his wallet. (In fact, he frequently misplaces things, including, on occasion, his car.) Thirty years ago, he created one of the world's first climate models, nick-named Model Zero, which he used to predict most of what has happened to the climate since. Sometimes he is referred to as the "father of global warming," and sometimes as the grandfather.

Hansen has now concluded, partly on the basis of his latest modeling efforts and partly on the basis of observations made by other scientists, that the threat of global warming is far greater than even he had suspected. Carbon dioxide isn't just approaching dangerous levels; it is already there. Unless immediate action is taken—including the shut-down of all the world's coal plants within the next two decades—the planet will be committed to change on a scale society won't be able to cope with. "This particular problem has become an emergency," Hansen said.

Hansen's revised calculations have prompted him to engage in activities—like marching on Washington—that aging government scientists don't usually go in for. Last September, he traveled to England to testify on behalf of anticoal activists who were arrested while climbing the smokestack of a power station to spray-paint a message to the prime minister. (They were acquitted.) Speaking before a congressional special committee last year, Hansen asserted that fossil-fuel companies were knowingly spreading misinformation about global warming and that their chairmen "should be tried for high crimes against humanity and nature." He has compared freight trains carrying coal to "death trains," and wrote to the head of the National Mining Association, who sent him a letter of complaint, that if the comparison "makes you uncomfortable, well, perhaps it should."

Hansen insists that his intent is not to be provocative but conservative: his only aim is to preserve the world as we know it. "The science is clear," he said, when it was his turn to address the protesters blocking the entrance to the Capitol Power Plant. "This is our one chance."

THE FIFTH OF SEVEN CHILDREN, Hansen grew up in Denison, Iowa, a small, sleepy town close to the western edge of the state. His father was a tenant farmer who, after the Second World War, went to work as a bartender. All the kids slept in two rooms. As soon as he was old enough, Hansen went to work too, delivering the *Omaha World-Herald*. When he was eighteen, he received a scholarship to attend the University of Iowa. It didn't cover housing, so he rented a room for twenty-five dollars a month and ate mostly cereal. He stayed on at the university to get a PhD in physics, writing his dissertation on the atmosphere of Venus. From there he went directly to the Goddard Institute for Space Studies—GISS, for short—where he took up the study of Venusian clouds.

By all accounts, including his own, Hansen was preoccupied by his research and not much interested in anything else. GISS's offices are a few blocks south of Columbia University, but when riots shut down the campus, in 1968, he barely noticed. At that point, GISS's computer was the fastest in the world, but it still had to be fed punch cards. "I was staying here late every night, reading in my decks of cards," Hansen recalled. In 1969, he left GISS for six months to study in the Netherlands. There he met his wife, Anniek, who is Dutch; the couple honeymooned in Florida, near Cape Canaveral, so they could watch an Apollo launch.

In 1973, the first Pioneer Venus mission was announced, and Hansen began designing an instrument—a polarimeter—to be carried on the orbiter. But soon his research interests began to shift earthward. A trio of chemists—they would later share a Nobel Prize—had discovered that chlorofluorocarbons and other man-made chemicals could break down the ozone layer. It had also become clear that greenhouse gases were rapidly building up in the atmosphere.

"We realized that we had a planet that was changing before our eyes, and that's more interesting," Hansen told me. The topic attracted him for much the same reason Venus's clouds had: there were new research questions to be answered. He decided to try to adapt a computer program that had been designed to forecast the weather to see if it could be used to look further into the future. What would happen to the earth if, for example, greenhouse gas levels were to double?

"He never worked on any topic thinking it might be any use for the world," Anniek told me. "He just wanted to figure out the scientific meaning of it."

When Hansen began his modeling work, there were good theoretical reasons for believing that increasing CO_2 levels would cause the world to warm, but little empirical evidence. Average global temperatures had risen in the nineteen-thirties and forties; then they had declined, in some regions, in the nineteen-fifties and sixties. A few years into his project, Hansen concluded that a new pattern was about to emerge. In 1981, he became the director of GISS. In a paper published that year in *Science,* he forecast that the following decade would be unusually warm. (That turned out to be the case.) In the same paper, he predicted that the nineteen-nineties would be warmer still. (That also turned out to be true.) Finally, he forecast that by the end of the twentieth century a global-warming signal would emerge from the "noise" of natural climate variability. (This, too, proved to be correct.)

Later, Hansen became even more specific. In 1990, he bet a roomful of scientists that that year, or one of the following two, would be the warmest on record. (Within nine months, he had won the bet.) In 1991, he predicted that, owing to the eruption of Mt. Pinatubo, in the Philippines, average global temperatures would drop and then, a few years later, recommence their upward climb, which was precisely what happened.

FROM EARLY ON, the significance of Hansen's insights was recognized by the scientific community. "The work that he did in the seventies, eighties, and nineties was absolutely groundbreaking," Spencer Weart, a physicist turned historian who has studied the efforts to understand climate change, told me. He added, "It does help to be right."

"I have a whole folder in my drawer labeled 'Canonical Papers,'" Michael Oppenheimer, a climate scientist at Princeton, said. "About half of them are Jim's."

Because of its implications for humanity, Hansen's work also attracted considerable attention from the world at large. His 1981 paper prompted the first front-page article on climate change that ran in *The New York Times*—"STUDY FINDS WARMING TREND THAT COULD RAISE SEA LEVELS," the headline read—and within a few years he was regularly being invited to testify before Congress. Still, Hansen says, he didn't imagine himself playing any role besides that of a research scientist. He is, he has written, "a poor communicator" and "not tactful."

"He's very shy," Ralph Cicerone, the president of the National Academy of Sciences, who has known Hansen for nearly forty years, told me. "And, as far as I can tell, he does not enjoy a lot of his public work."

"Jim doesn't really like to look at anyone," Anniek Hansen told me. "I say, 'Just look at them!'"

Throughout the nineteen-eighties and nineties, the evidence of climate change—and its potential hazards—continued to grow. Hansen kept expecting the political system to respond. This, after all, was what had happened with the ozone problem. Proof that chlorofluorocarbons were destroying the ozone layer came in 1985, when British scientists discovered that an ozone "hole" had opened up over Antarctica. The crisis was resolved—or, at least, prevented from growing worse— by an international treaty phasing out chlorofluorocarbons that was ratified in 1987.

"At first, Jim's work didn't take an activist bent at all," the writer Bill McKibben, who has followed Hansen's career for more than twenty years and helped organize the anti-coal protest in D.C., told me. "I think he thought, as did I, 'If we get this set of facts out in front of everybody, they're so powerful—overwhelming—that people will do what needs to be done.' Of course, that was naive on both our parts."

As recently as the George W. Bush administration, Hansen was still operating as if getting the right facts in front of the right people would be enough. In 2001, he was invited to speak to Vice President

Dick Cheney and other high-level administration officials. For the meeting, he prepared a detailed presentation titled "The Force Underlying Climate Change." In 2003, he was invited to Washington again, to meet with the head of the Council on Environmental Quality at the White House. This time, he offered a presentation on what ice-core records show about the sensitivity of the climate to changes in greenhouse gas concentrations. But by 2004 the administration had dropped any pretense that it was interested in the facts about climate change. That year, NASA, reportedly at the behest of the White House, insisted that all communications between GISS scientists and the outside world be routed through political appointees at the agency. The following year, the administration prevented GISS from posting its monthly temperature data on its website, ostensibly on the grounds that proper protocols had not been followed. (The data showed that 2005 was likely to be the warmest year on record.) Hansen was also told that he couldn't grant a routine interview to National Public Radio. When he spoke out about the restrictions, scientists at other federal agencies complained that they were being similarly treated and a new term was invented: government scientists, it was said, were being "Hansenized."

"He had been waiting all this time for global warming to become the issue that ozone was," Anniek Hansen told me. "And he's very patient. And he just kept on working and publishing, thinking that someone would do something." She went on, "He started speaking out, not because he thinks he's good at it, not because he enjoys it, but because of necessity."

"When Jim makes up his mind, he pursues whatever conclusion he has to the end point," Michael Oppenheimer said. "And he's made up his mind that you have to pull out all the stops at this point, and that all his scientific efforts would come to naught if he didn't also involve himself in political action." Starting in 2007, Hansen began writing to world leaders, including Prime Minister Gordon Brown, of Britain, and Yasuo Fukuda, then the prime minister of Japan. In December 2008, he composed a personal appeal to Barack and Michelle Obama.

"A stark scientific conclusion, that we must reduce greenhouse gases below present amounts to preserve nature and humanity, has become clear," Hansen wrote. "It is still feasible to avert climate disasters, but only if policies are consistent with what science indicates to be required." Hansen gave the letter to Obama's chief science adviser, John Holdren, with whom he is friendly, and Holdren, he says, promised to deliver it. But Hansen never heard back, and by the spring he had begun to lose faith in the new administration. (In an email, Holdren said that he could not discuss "what I have or haven't given or said to the President.")

"I had had hopes that Obama understood the reality of the issue and would seize the opportunity to marry the energy and climate and national security issues and make a very strong program," Hansen told me. "Maybe he still will, but I'm getting bad feelings about it."

THERE ARE LOTS of ways to lose an audience with a discussion of global warming, and new ones, it seems, are being discovered all the time. As well as anyone, Hansen ought to know this; still, he persists in trying to make contact. He frequently gives public lectures; just in the past few months, he has spoken to Native Americans in Washington, D.C.; college students at Dartmouth; high school students in Copenhagen; concerned citizens, including King Harald, in Oslo; renewable-energy enthusiasts in Milwaukee; folk music fans in Beacon, New York; and public health professionals in Manhattan.

In April, I met up with Hansen at the state capitol in Concord, New Hampshire, where he had been invited to speak by local anti-coal activists. There had been only a couple of days to publicize the event; nevertheless, more than two hundred and fifty people showed up. I asked a woman from the town of Ossipee why she had come. "It's a once-in-a-lifetime opportunity to hear bad news straight from the horse's mouth," she said. For the event, Hansen had, as usual, prepared a PowerPoint presentation. It was projected onto a screen beside a faded portrait of George Washington. The first slide gave the title of the talk, "The Climate Threat to the Planet," along with the disclaimer "Any statements relating to policy are personal opinion."

Hansen likes to begin his talk with a highly compressed but still perilously long discussion of climate history, beginning in the early Eocene, some fifty million years ago. At that point, CO_2 levels were high and, as Hansen noted, the world was very warm: there was practically no ice on the planet, and palm trees grew in the Arctic. Then CO_2 levels began to fall. No one is entirely sure why, but one possible cause has to do with weathering processes that, over many millennia, allow carbon dioxide from the air to get bound up in limestone. As CO_2 levels declined, the planet grew cooler; Hansen flashed some slides on the screen, which showed that, between fifty million and thirty-five million years ago, deep-ocean temperatures dropped by more than ten degrees. Eventually, around thirty-four million years ago, temperatures sank low enough that glaciers began to form on Antarctica. By around three million years ago—perhaps earlier—permanent ice sheets had begun to form in the Northern Hemisphere as well. Then, about two million years ago, the world entered a period of recurring glaciations. During the ice ages—the most recent one ended about twelve thousand years ago—CO_2 levels dropped even further.

What is now happening, Hansen explained to the group in New Hampshire, is that climate history is being run in reverse and at high speed, like a cassette tape on rewind. Carbon dioxide is being pumped into the air some ten thousand times faster than natural weathering processes can remove it.

"So humans now are in charge of atmospheric composition," Hansen said. Then he corrected himself: "Well, we're determining it, whether we're in charge or not." Among the many risks of running the system backward is that the ice sheets formed on the way forward will start to disintegrate. Once it begins, this process is likely to be self-reinforcing. "If we burn all the fossil fuels and put all that CO_2 into the atmosphere, we will be sending the planet back to the ice-free state," Hansen said. "It will take a while to get there—ice sheets don't melt instantaneously—but that's what we will be doing. And if you melt all the ice, sea levels will go up two hundred and fifty feet. So you can't do that without producing a different planet."

• • •

THERE'S NO PRECISE term for the level of CO_2 that will assure a climate disaster; the best that scientists and policymakers have been able to come up with is the phrase "dangerous anthropogenic interference," or DAI. Most official discussions have been premised on the notion that DAI will not be reached until CO_2 levels hit four hundred and fifty parts per million. Hansen, however, has concluded that the threshold for DAI is much lower.

"The bad news is that it's become clear that the dangerous amount of carbon dioxide is no more than three hundred and fifty parts per million," he told the crowd in Concord. The really bad news is that CO_2 levels have already reached three hundred and eighty-five parts per million. For the ten thousand years prior to the industrial revolution, carbon dioxide levels were about two hundred and eighty parts per million, and if current emissions trends continue they will reach four hundred and fifty parts by around 2035.

Once you accept that CO_2 levels are already too high, it's obvious, Hansen argues, what needs to be done. He displayed a chart of known fossil-fuel reserves represented in terms of their carbon content. There was a short bar for oil, a shorter bar for natural gas, and a tall bar for coal.

"We've already used about half of the oil," he observed. "And we're going to use all of the oil and natural gas that's easily available. It's owned by Russia and Saudi Arabia, and we can't tell them not to sell it. So, if you look at the size of these fossil-fuel reservoirs, it becomes very clear. The only way we can constrain the amount of carbon dioxide in the atmosphere is to cut off the coal source, by saying either we will leave the coal in the ground or we will burn it only at power plants that actually capture the CO_2." Such power plants are often referred to as "clean coal plants." Although there has been a great deal of talk about them lately, at this point there are no clean-coal plants in commercial operation, and, for a combination of technological and economic reasons, it's not clear that there ever will be.

Hansen continued, "If we had a moratorium on any new coal plants and phased out existing ones over the next twenty years, we could get back to three hundred and fifty parts per million within several decades." Reforestation, for example, if practiced on a massive scale, could begin to draw global CO_2 levels down, Hansen says, "so it's technically feasible." But "it requires us to take action promptly."

Coincidentally, that afternoon a vote was scheduled in the New Hampshire state legislature on a proposal involving the state's largest coal-fired power plant, the Merrimack Station, in the town of Bow. The station's owner was planning to spend several hundred million dollars to reduce mercury emissions from the plant—a cost that it planned to pass on to ratepayers. Hansen, who said he thought the plant should simply be shut, called the plan a "terrible waste of money." A lawmaker sympathetic to this view had introduced a bill calling for more study of the project, but, as several people who came up to speak to Hansen after his talk explained, it was opposed by the state's construction unions and seemed headed for defeat. (Less than an hour later, the bill was rejected in committee by a unanimous vote.)

"I assume you're used to telling policymakers the truth and then having them ignore you," one man said to Hansen.

Hansen smiled ruefully. "You're right."

IN SCIENTIFIC CIRCLES, worries about DAI are widespread. During the past few years, researchers around the world have noticed a disturbing trend: the planet is changing faster than had been anticipated. Antarctica, for example, had not been expected to show a net loss of ice for another century, but recent studies indicate that the continent's massive ice sheets are already shrinking. At the other end of the globe, the Arctic ice cap has been melting at a shocking rate; the extent of the summer ice is now only a little more than half of what it was just forty years ago. Meanwhile, scientists have found that the arid zones that circle the globe north and south of the tropics have been expanding more rapidly than computer models had predicted. This expansion of the subtropics means that highly populated areas, including the American Southwest and the Mediterranean basin, are likely to suffer more and more frequent droughts.

"Certainly, I think the shrinking of the Arctic ice cap made a very strong impression on a lot of scientists," Spencer Weart, the physicist, told me. "And these things keep popping up. You think, What, another one? Another one? They're almost all in the wrong direction, in the direction of making the change worse and faster."

"In nearly all areas, the developments are occurring more quickly than had been assumed," Hans Joachim Schellnhuber, the head of Germany's Potsdam Institute for Climate Impact Research, recently observed. "We are on our way to a destabilization of the world climate that has advanced much further than most people or their governments realize."

Obama's science adviser, John Holdren, a physicist on leave from Harvard, has said that he believes "any reasonably comprehensive and up-to-date look at the evidence makes clear that civilization has already generated dangerous anthropogenic interference in the climate system."

There is also broad agreement among scientists that coal represents the most serious threat to the climate. Coal now provides half the electricity in the United States. In China, that figure is closer to 80 percent, and a new coal-fired power plant comes online every week or two. As oil supplies dwindle, there will still be plenty of coal, which could be—and in some places already is being, converted into a very dirty liquid fuel. Before Steven Chu, a Nobel Prize–winning physicist, was appointed to his current post as energy secretary, he said in a speech, "There's enough carbon in the ground to really cook us. Coal is my worst nightmare." (These are lines that Hansen is fond of invoking.) A couple of months ago, seven prominent climate scientists from Australia wrote an open letter to the owners of that country's major utility companies urging that "no new coal-fired power stations, except ones that have ZERO emissions," be built. They also recommended an "urgent program" to phase out old plants.

"The unfortunate reality is that genuine action on climate change will require that existing coal-fired power stations cease to operate in the near future," the group wrote.

But if Hansen's anxieties about DAI and coal are broadly shared, he is still, among climate scientists, an outlier. "Almost everyone in the scientific community is prepared to say that if we don't do something now to reverse the direction we're going in, we either already are or will very, very soon be in the danger zone," Naomi Oreskes, a historian of science and a provost at the University of California at San Diego, told me. "But Hansen talks in stronger terms. He's using adjectives. He has started to speak in moral terms, and that always makes scientists uncomfortable."

Hansen is also increasingly isolated among climate activists. "I view Jim Hansen as heroic as a scientist," Eileen Claussen, the president of the Pew Center on Global Climate Change, said. "He was there at the beginning, he's faced all kinds of pressures politically, and he's done a terrific job, I think, of keeping focused. But I wish he would stick to what he really knows. Because I don't think he has a realistic view of what is politically possible, or what the best policies would be to deal with this problem."

In Washington, the only approach to limiting emissions that is seen as having any chance of being enacted is a so-called cap-and-trade system. Under such a system, the government would set an overall cap for CO_2 emissions, then allocate allowances to major emitters, like power plants and oil refineries, that could be traded on a carbon market. In theory, at least, the system would discourage fossil-fuel use by making emitters pay for what they are putting out. But to the extent that such a system has been tried, by the members of the European Union, its results so far are inconclusive, and Hansen argues that it is essentially a sham. (He recently referred to it as "the Temple of Doom.") What is required, he insists, is a direct tax on carbon emissions. The tax should be significant at the start—equivalent to roughly a dollar per gallon for gasoline—and then should grow steeper over time. The revenues from the tax, he believes, ought to be distributed back to Americans on a per capita basis, so that households that use less energy would actually make money, even as those that use more would find it increasingly expensive to do so.

"The only defense of this monstrous absurdity that I have heard," Hansen wrote a few weeks ago, referring to a cap-and-trade system, "is 'Well, you are right, it's no good, but the train has left the station.' If the train has left, it had better be derailed soon or the planet, and all of us, will be in deep doo-doo."

GISS'S HEADQUARTERS, at 112th Street and Broadway, sits above Tom's Restaurant, the diner made famous by *Seinfeld* and Suzanne Vega. Hansen has occupied the same office, on the seventh floor, since he became the director of the institute, almost three decades ago. One day last month, I went to visit him there. Hansen told me that he had been trying to computerize his old files; still, the most striking thing about the spacious office, which is largely taken up by three wooden tables, is that every available surface is covered with stacks of paper.

During the week, Hansen lives in an apartment just a few blocks from his office, but on weekends he and Anniek frequently go to an eighteenth-century house that they own in Bucks County, Pennsylvania, and their son and daughter, who have children of their own, come to visit. Hansen dotes on his grandchildren—in many hours of conversation with me, just about the only time that he spoke with unalloyed enthusiasm was when he discussed planting trees with them this spring—and he claims they are the major reason for his activism. "I decided that I didn't want my grandchildren to say, 'Opa understood what was happening, but he didn't make it clear,'" he explained.

The day that I visited Hansen's office, the House Energy and Commerce Committee was beginning its markup of a cap-and-trade bill cosponsored by the committee's chairman, Henry Waxman, of California. The bill—the American Clean Energy and Security Act—has the stated goal of cutting the country's carbon emissions by 17 percent by 2020. It is the most significant piece of climate legislation to make it this far in the House. Hansen pointed out that the bill explicitly allows for the construction of new coal plants and predicted that it would, if passed, prove close to meaningless. He said that he thought it would probably be best if the bill failed, so that Congress could "come back and do it more sensibly."

I said that if the bill failed I thought it was more likely Congress would let the issue drop, and that was one reason most of the country's major environmental groups were backing it.

"This is just stupidity on the part of environmental organizations in Washington," Hansen said. "The fact that some of these organizations have become part of the Washington 'go along, get along' establishment is very unfortunate."

Hansen argues that politicians willfully misunderstand climate science; it could be argued that Hansen just as willfully misunderstands politics. In order to stabilize carbon dioxide levels in the atmosphere, annual global emissions would have to be cut by something on the order of three-quarters. In order to draw them down, agricultural and forestry practices would have to change dramatically as well. So far, at least, there is no evidence that any nation is willing to take anything approaching the necessary steps. On the contrary, almost all the trend lines point in the opposite direction. Just because the world desperately needs a solution that satisfies both the scientific and the political constraints doesn't mean one necessarily exists.

For his part, Hansen argues that while the laws of geophysics are immutable, those of society are ours to determine. When I said that it didn't seem feasible to expect the United States to give up its coal plants, he responded, "We can point to other countries being 50 percent more energy-efficient than we are. We're getting 50 percent of our electricity from coal. That alone should provide a pretty strong argument."

Then what about China and India?

Both countries are likely to suffer very severely from dramatic climate change, he said. "They're going to recognize that. In fact, they already are beginning to recognize that."

"It's not unrealistic," he went on. "But the policies have to push us in that direction. And, as long as we let the politicians and the people who are supporting them continue to set the rules, such that 'business as usual' continues, or small tweaks to 'business as usual,' then it is unrealistic. So we have to change the rules." He said that he was thinking of attending another demonstration soon, in West Virginia coal country.

Originally published in *The New Yorker,*
June 29, 2009.

Update: The proportion of the U.S.'s electricity generated from coal has dropped to less than 20 percent, as power generation from wind, solar, and—most significantly—natural gas has increased. Meanwhile, the country's electricity demand, which had flattened out, is once again climbing.

THE WEIGHT OF THE WORLD

Profile of Christiana Figueres,
One of the Architects of the Paris Agreement

THE UNITED NATIONS Framework Convention on Climate
Change, or UNFCCC, has by now been ratified by a hundred and
ninety-five countries, which, depending on how you count, represents
either all the countries in the world or all the countries and then
some. Every year, the treaty stipulates, the signatories have to hold a
meeting—a gathering that's known as a COP, short for Conference
of the Parties. The third COP produced the Kyoto Protocol, which, in
turn, gave rise to another mandatory gathering, a MOP, or Meeting of
the Parties. The seventeenth COP, which coincided with the seventh
MOP, took place in South Africa. There it was decided that the work
of previous COPs and MOPs had been inadequate, and a new group
was formed: the Ad Hoc Working Group on the Durban Platform for
Enhanced Action, usually referred to as the ADP. The ADP subsequently split into ADP-1 and ADP-2, each of which held meetings of
its own. The purpose of the UNFCCC and of the many negotiating
sessions and working groups and protocols it has spun off over the
years is to prevent "dangerous anthropogenic interference with the climate system." In climate circles, this is usually shortened to DAI. In
plain English, it means global collapse.

The Framework Convention on Climate Change is overseen by an
organization known as the Secretariat, which is led by a Costa Rican
named Christiana Figueres. Figueres is five feet tall, with short brown
hair and strikingly different-colored eyes—one blue and one hazel. In

contrast to most diplomats, who cultivate an air of professional reserve, Figueres is emotive to the point of disarming—"a mini-volcano" is how one of her aides described her to me. She laughs frequently—a hearty, ha-ha-ha chortle—and weeps almost as often. "I walk around with Kleenex," another aide told me.

Figueres, who is fifty-nine, is an avid runner—the first time I met her, she was hobbling around with blisters acquired from a half marathon—and an uninhibited dancer. Last fall, when her office was preparing for the twentieth COP, which was held in Lima, she and some of her assistants secretly practiced a routine set to Beyoncé's "Move Your Body." At a meeting of the Secretariat staff, which numbers more than five hundred, they ripped off their jackets and started to jump, jump, jump.

Figueres works out of a spacious office in Bonn, in a building that used to belong to the German parliament. On the wall by her desk there's a framed motto that reads, "Impossible is not a fact, it is an attitude." On another wall there's a poster showing the Statue of Liberty waist-high in water, and on a third a black-and-white photograph of Figueres's father, José, who led the Costa Rican revolution of 1948. He served as president of the country three times, pushed through sweeping political and social reforms, and abolished Costa Rica's army as a stay against dictatorship. Figueres grew up partly in the President's House and partly on her father's farm, which he called La Lucha sin Fin—"the struggle without end."

"I'm very comfortable with the word *revolution*," Figueres told me. "In my experience, revolutions have been very positive."

Of all the jobs in the world, Figueres's may possess the very highest ratio of responsibility (preventing global collapse) to authority (practically none). The role entails convincing a hundred and ninety-five countries—many of which rely on selling fossil fuels for their national income and almost all of which depend on burning them for the bulk of their energy—that giving up such fuels is a good idea. When Figueres took over the Secretariat, in 2010, there were lots of people who thought the job so thankless that it ought to be abolished. This was in the aftermath of the fifteenth COP, held in Copenhagen, which had been expected to yield a historic agreement but ended in anger and recrimination.

Figueres and her team have spent the years since Copenhagen try-ing to learn from its mistakes. How well they have done so will become apparent three months from now, when world leaders meet for this year's COP—the twenty-first—in Paris. Like Copenhagen, Paris is being billed as a historic event—"our last hope," in the words of Fatih Birol, the incoming director of the International Energy Agency—and, again, expectations are running high. "We are duty-bound to suc-ceed," France's president, François Hollande, has declared.

The danger of high expectations, of course, is that they can be all the more devastatingly dashed. Figueres, who is well aware of this, is doing her best to raise them further, on the theory that the best way to make something happen is to convince people that it is going to hap-pen. "I have not met a single human being who's motivated by bad news," she told me. "Not a single human being."

TO UNDERSTAND HOW the fate of the planet came to be en-trusted to a corps of mostly anonymous, midlevel diplomats, you have to go back to the nineteen-eighties, when the world confronted its first atmospheric crisis. That crisis, the so-called ozone hole, was the product of chemicals known as chlorofluorocarbons, or CFCs. When they were invented, in the nineteen-twenties, CFCs were hailed as miracle compounds, safe alternatives to the toxic gases used as early refrigerants. Lots of additional uses were found for CFCs before it was discovered that the chemicals had the nasty effect of breaking down stratospheric ozone, which protects the earth from ultraviolet radiation. (F. Sherwood Rowland, a chemist who shared a Nobel Prize for this discovery, once reportedly came home from his lab and told his wife, "The work is going well, but it looks like it might be the end of the world.") A global treaty—the Vienna Con-vention for the Protection of the Ozone Layer—was signed in 1985 and sent to the U.S. Senate by President Ronald Reagan, who called for its "expeditious ratification." This broadly worded "framework" was soon followed by the Montreal Protocol, which called for drastic cuts in CFC usage.

The Montreal Protocol, which has been revised a half-dozen times, mainly in response to new scientific data, averted a dystopian future filled with skin cancer and cataracts. (If you're sitting in the sun right now, in a roundabout way you can thank the Montreal Protocol.) Kofi Annan, the former secretary-general of the UN, has labeled it "perhaps the single most successful international agreement to date."

When scientists first sounded the alarm about carbon emissions, it seemed logical to try to follow the Montreal template. In 1988, the UN General Assembly adopted a resolution declaring climate change to be a "common concern of mankind." The following year, talks began on what was to become the Framework Convention.

The ozone treaty had divided the world into two blocs: high CFC users, like the United States, and low users, like Bangladesh. High users, who were largely responsible for the problem, were expected to act first, low users later. The same high-low distinction held for climate change; some countries had contributed a great deal to the problem, others very little. But almost immediately the blocs fractured into sub-blocs. Oil-producing states, like Saudi Arabia, split with low-lying, easily inundated nations, like the Maldives. Rapidly industrial-izing countries, like India, saw their interests as very different from those of what are officially known as Least Developed Countries, like Ethiopia. The European Union wanted a treaty with strict targets and timetables for reducing carbon emissions. The United States—at that point the world's largest emitter—refused even to consider such targets. On the eve of what was supposed to be the final negotiating session on the Framework Convention, the working draft of the docu-ment, according to one participant, resembled a "compilation of con-tradictory positions more than a recognizable legal instrument."

The convention was rescued, at a price. The final version of the treaty, presented in Rio in 1992, called for the "stabilization of green-house gas concentrations in the atmosphere at a level that would pre-vent dangerous anthropogenic interference with the climate system." But it left virtually all decisions about how this was to be accomplished to future negotiations. Also left unresolved was how those decisions were to be reached: the convention provided no rules for voting, though it noted that such rules ought to be adopted.

This constitutive vagueness has troubled climate negotiations ever since. The parties to the convention have never managed to agree on rules for voting, meaning that every decision must—in theory, at least—be arrived at by consensus. And while a few countries have cut their CO_2 output since the convention was signed, globally emissions have soared, from about twenty-two billion metric tons of carbon dioxide a year in the early nineteen-nineties to more than thirty-five billion metric tons today.

Figueres lives about five miles from downtown Bonn, on the opposite side of the Rhine, in the town of Königswinter. She owns a Prius but usually takes the tram to work. One evening this spring, I rode the tram home with her. She had spent the previous night in Munich and was dragging a rolling suitcase behind her. On the walk from the tram stop to her apartment, she dropped by a market to buy food for dinner. She had no shopping bag, so she decided to carry the groceries home in her suitcase.

"I'm not Alice in Wonderland," she told me, once we got upstairs. "You and I are sitting here, in this gorgeous apartment, enjoying this fantastic privilege, because of fossil fuels." Figueres, who is separated from her husband, has two grown daughters, one of whom works in New York, the other in Panama. Her apartment is decorated with vividly colored paintings by Central American artists, and it looks directly onto the Rhine, which, on this particular evening, was untrafficked except for an occasional coal barge.

At the time of my visit, Figueres was preparing for a trip to Saudi Arabia. Over drinks on her balcony, she described what it had been like working with the Saudis at climate negotiations when she herself was a delegate, representing Costa Rica. "They would throw a wrench in here and get out of that room in which the issue was A, then appear over in this other room, in which it was a completely unrelated issue, throw a wrench in there, and disappear," she recalled. "I would stand there with my mouth open. I would go, These guys are brilliant.

"The Saudis are sitting on a vast reserve of very cheap oil," she continued. "Can you blame them for trying to protect that resource and that income for as long as they can? I don't blame them. It's very understandable. Let's do a thought experiment. I come from a country that has only hydro and wind as power resources. If I had been born in a country with fossil-fuel reserves, would I have a different opinion about what's good for the world? Maybe. Very likely, in fact.

"I don't want to put people into a black box and say, 'You're the culprits,' and point a blaming finger. It just helps absolutely nothing. Call it my anthropological training. Call it whatever. But I always want to understand: What is behind all of this?"

Growing up in Costa Rica, Figueres was sent to the Humboldt School, in San José, where she learned to speak fluent German. For college, she came to the United States—to Swarthmore—where she studied anthropology. Then she spent a year working with the Bribri, an Indigenous people who live in Costa Rica's Talamanca Mountains. The village had no electricity or running water.

"I have no problem sitting on the floor, sipping hot water from a dirty cup," she told me. "I also have no problem sitting next to Prince Charles." Figueres had brought along a camera to document the Bribris' lives. She discovered that they loved to see photographs of themselves, and so every few months she would trek out of the village, by foot and by donkey, to get the pictures developed. Once, she also brought back a postcard showing New York City at night: "I thought, 'Let's see how they interpret this.' So I just showed them the photograph, and I said, 'What is this?'

"'Ah,' they said. 'All the little stars of heaven in rows!' What a beautiful interpretation. They had no concept of what a lit city was. The only light they had seen at night was the stars. And then, all of a sudden, all these little stars were in rows! Now, funnily enough, I think about that response almost daily. Because my feeling is that all the little stars are aligning themselves in a different sense."

It is Figueres's contention that all the nations of the world are now working in good faith to try to reach a climate agreement, and that includes Saudi Arabia. She cited her invitation from the country as a sign of its new, more "constructive" approach. On her trip, she wanted to be careful to adhere to the nation's strict dress code, so she had had her secretary call to ask what was expected of her. "I know that in Riyadh I need to wear a burka," she told me. "Elsewhere, if they want me to wear an abaya I'll wear an abaya."

She also wanted to be mindful of the Saudis' linguistic requirements. "They don't like the term *low carbon*," she explained. "They don't like the term *decarbonization,* because for them that points the finger directly at them. They would rather use the term *low emissions.*" This spreads the blame for global warming to other greenhouse gases, like nitrous oxide and hydrofluorocarbons. (In an unfortunate irony, hydrofluorocarbons, which, molecule for molecule, trap far more heat than CO_2, were specifically engineered to replace ozone-depleting CFCs.)

"Well, frankly, I sometimes do talk about 'decarbonization,'" Figueres went on. "But certainly I won't talk about decarbonization when I'm in Saudi Arabia, because I understand that is very threatening to them. Why would I want to threaten them? I need them on my side. The best thing that could happen to me would be that Saudi Arabia says, 'You know what? With all the money that we have, we're going to invest in the best technology in concentrated solar power.'"

IN THE LEAD-UP to Paris, each country has been asked to submit a plan outlining how and by how much it will reduce its carbon output—or, to use the Saudis' preferred term, its emissions. The plans are known as "intended nationally determined contributions"—in UN-speak, INDCs. The whole approach has been labeled "bottom up," which, by implication, makes previous efforts to cut carbon—in particular, the Kyoto Protocol—"top down."

Drafted in 1997, Kyoto represented the first and, as yet, the only time that the parties managed to fill in some of the Framework Convention's many blanks. People who attended that COP still remember it as a kind of endurance test for the soul. Figueres, who was there with the Costa Rican delegation, described it as "an absolutely harrowing experience." Kyoto imposed specific targets on roughly forty countries of the Global North (not all of which, of course, are actually in the north). The targets varied from country to country; the nations of the European Union, for instance, were, collectively, supposed to cut their emissions by 8 percent, while the United States was supposed to cut them by 7 percent. (This was against a baseline of 1990.) Canada was expected to reduce its emissions by 6 percent. Australia's target allowed its emissions to grow, but not beyond 8 percent.

Countries in the Global South were not given targets, on the theory that it would be unfair to ask them to reduce their already relatively small output. (Saudi Arabia, part of this second group, tried to scuttle the agreement in advance by demanding that the text be circulated six months before the final negotiating session.) It was the United States that helped rescue the protocol—Vice President Al Gore flew to Kyoto when the talks appeared to be foundering—and it was also the United States that very nearly killed it. The Senate refused to ratify the treaty, and shortly after George W. Bush entered the White House, in 2001, he announced that his administration would not abide by its terms.

"Kyoto is dead" is how Condoleezza Rice, Bush's national security adviser, put it. In fact, the treaty survived, but in a zombielike state. The United States ignored it. The Canadians blew past their target and, midway through the period covered by Kyoto, withdrew from the agreement. Only the Europeans really took their goal seriously, not only meeting it but exceeding it.

Meanwhile, as Kyoto shambled on, the horizon receded. In the mid-nineties, China was emitting nearly a billion metric tons of carbon a year. By the mid-aughts, its output was twice that amount. In 2005, China surpassed the United States as the world's largest emitter on an annual basis. (The United States still holds first place in terms of cumulative emissions.) Nowadays, China's per capita emissions are as high as western Europe's (though not nearly as high as those in the United States). The more than a thousand new coal-fired power plants that went up from Guangdong to Xinjiang made Dutch wind turbines and German solar farms seem increasingly irrelevant; all of Europe's cuts were effectively canceled out by a few months' worth of emissions growth in China. Scientists warned that the world was on track for an average global temperature rise of four degrees Celsius (more than seven degrees Fahrenheit) by the end of this century. Such a temperature increase, they predicted, would transform the globe into a patchwork of drowned cities, desertifying croplands, and collapsing ecosystems. As a report from the World Bank noted, it's not clear "that adaptation to a 4°C world is possible."

It was to get off this path that negotiators met in Copenhagen in 2009. The new plan was supposed to establish stricter targets and extend them to more countries, including China. Instead, what emerged from the session was yet another prolegomenon to future negotiations, brokered at the very last minute—and over the objections of many other world leaders—by President Barack Obama. Known as the Copenhagen Accord, the document—a sort of climate wish list— identified a temperature rise of two degrees Celsius as the danger point for the planet. It also promised funding to help poor countries affected by warming. This funding is supposed to amount to a hundred billion dollars a year.

FIGUERES SPENDS MUCH of her time traveling around the globe, meeting with anyone she thinks might advance the cause. A few weeks after she visited Saudi Arabia, she went to London, where she spoke to, among others, Bill Gates, Richard Branson, and Al Gore. "In my opinion, Christiana has done a terrific job under excruciatingly difficult circumstances," Gore told me. From London, she flew to New York for three more days of back-to-back meetings.

One morning began with a breakfast at Citigroup's headquarters, on Park Avenue. Seated around the table were the New York State comptroller, the chief investment officer of Connecticut's pension funds, and representatives of several major investment firms. Figueres began by assuring the group that the negotiations leading to Paris were "still on track." Then she turned her attention to money.

"Where capital goes over the next fifteen years is going to decide whether we're actually able to address climate change and what kind of a century we are going to have," she said. She urged all those present to take this into account when making their own investment decisions, and to do so publicly: "What we truly need is to create a 'surround sound' where, no matter what sector you turn to, there is a signal saying, 'Folks, we are moving toward a low-carbon economy. It is irreversible; it is unstoppable. So get on the bandwagon.'"

The debate over what to do—or not to do—about global warming has always been, at its core, an economic one. Since the start of the industrial revolution, growth has been accompanied—indeed, made possible—by rising emissions. Hence the reluctance of most nations to commit to cutting carbon. But what if growth and emissions could be uncoupled?

In some parts of Europe, what has been called "conscious uncoupling" is already well along. Sweden, one of the few countries that tax carbon, has reduced its emissions by about 23 percent in the past twenty-five years. During that same period, its economy has grown by more than 55 percent. Last year, perhaps for the first time since the invention of the steam engine, global emissions remained flat even as the global economy grew, by about 3 percent.

Figueres maintains that global uncoupling is not only possible but obligatory. "We frankly don't have an option," she told me. "Because there are two things that are absolutely key to being able to feed, house, and educate the two billion more family members who will be joining us. You have to continue to grow. And, particularly, developing countries need to continue to grow. But the other sine qua non condition is that you can't continue to grow greenhouse gases, because that kills the possibility of growth. So, since you have those two constant constraints—you have to grow GDP, but you cannot grow GHGs— what option do you have?"

The day Figueres met with investors at Citigroup, China submitted its emissions plan, or INDC. The plan elaborated on the country's pledge to "peak" its emissions by 2030. That pledge, first made public in November, was part of a deal announced jointly with the United States. At the time, it was hailed as a breakthrough, as China had previously resisted making any commitment to capping its emissions, ever. *The Washington Post* labeled the announcement a "landmark"; *Time* called it "historic."

It was "huge, absolutely huge," Figueres told me. In its INDC, China said that it would make its "best efforts" to cap its emissions earlier than 2030. It also pledged to increase its share of energy produced from "non–fossil fuels" to 20 percent. (As one commentator pointed out, to fulfill the latter promise China will have to add enough solar, wind, or nuclear generating capacity to power the U.S.'s entire electrical grid.)

Iceland, Serbia, and South Korea also submitted their plans that day. In its submission, South Korea, the world's eleventh-largest emitter, committed to only trivial reductions, saying it could not go further because its economy is so heavily based on manufacturing. Analysts criticized the South Korean plan as essentially meaningless. Nevertheless, Figueres included the country in her Twitter feed. "Thx Republic of Korea," she tweeted after leaving Citigroup. She was headed over to the UN to meet with Secretary-General Ban Ki-moon.

WHEN CLIMATE SCIENTISTS talk about carbon, they usually do so in terms of parts per million. During the last ice age, when much of North America was covered in glaciers a mile thick, carbon dioxide levels in the atmosphere were around a hundred and eighty parts per million. For the ten thousand years leading up to the industrial revolution, they hovered around two hundred and eighty parts per million. By 1992, when the Framework Convention was drafted, they had reached three hundred and fifty parts per million. As MOP followed COP, carbon dioxide levels kept rising. This spring, they topped four hundred parts per million.

Owing to the CO_2 that's been pumped into the air so far, average global temperatures have risen by about 0.85 degrees Celsius (1.5 degrees Fahrenheit). This relatively small increase has produced some very large effects: almost half of the permanent Arctic ice cap has melted away, millions of acres' worth of trees in the American West have died from warming-related pest infestations, and some of West

Antarctica's major glaciers, containing tens of thousands of cubic miles of ice, have started to disintegrate.

To hold warming to less than two degrees Celsius, the best estimates available suggest that total emissions will have to be kept under a trillion tons of carbon. The world has already consumed around two-thirds of this budget. If current trends continue, the last third will be used up within the next few decades. What's fundamentally at issue in Paris—although the matter is never stated this baldly, because, if it were, the conference might as well be called off—is who should be allowed to emit the tons that remain.

One approach would be to assign these tons on the basis of aggregate emissions. Under this approach, very large emitters, like the United States, might not get any of the dwindling slice, on the ground that they've already gobbled up so much. Another way to allocate emissions would be to grant everyone on the planet an equal share of what's left; in that case, the United States would still have to radically reduce its emissions, but not all the way down to zero. A third approach would be to focus on efficiency. It's expensive to shutter power plants and factories that have already been built. But, as the cost of renewable energy declines, it may be cheaper to, say, put up solar panels than to construct a new coal plant. If growth can truly be decoupled from emissions, then poor countries shouldn't need to burn through lots of carbon in order to become wealthy.

The INDCs obviate the need to agree on a single approach, or even to disagree. Each country brings its own proposal to the table, as if to a planet-wide potluck.

The United States' plan is a peculiarly American confection. It consists of steps that the Obama administration can take without the support of Congress, whose Republican leadership regards global warming as some kind of liberal plot. (The House Speaker, John Boehner, once called the idea that carbon dioxide emissions are harmful "almost comical.") New rules for cars and light trucks, issued by the Department of Transportation, are supposed to raise the fuel efficiency of the average vehicle to 54.5 miles per gallon over the next decade, and new rules on power-plant emissions, recently finalized by the Environmental

Protection Agency, are expected to force the closure of dozens of the country's least efficient coal-fired plants.

According to the administration's calculations, these new rules, coupled with stricter energy-efficiency standards for equipment and appliances, will lower U.S. emissions by at least 26 percent by 2025. (This is against a baseline of 2005.) President Obama has said that this is an "ambitious goal, but it's an achievable goal." Still, it will leave the world on track to burn through its two-degree budget within a matter of decades.

WHILE FIGUERES WAS in New York, she attended a daylong meeting on climate change that had been called by the president of the UN General Assembly, Sam Kutesa, of Uganda. To such events, Figueres usually wears a Hillary Clinton–esque outfit consisting of black slacks, black pumps, and a short, colorful jacket. (On this particular day, the jacket was teal.) As she made her way into the General Assembly building, she ran into China's lead negotiator on climate issues, Xie Zhenhua. She joked to Xie that she was sure he had all the solutions in his pocket.

"Oh, no, no, no," Xie answered, laughing nervously, once the joke had been conveyed to him by his translator.

"I believe that, under your leadership, success is a must," he told Figueres.

"Under everyone's leadership," she responded.

The meeting began with speeches by various dignitaries, including Anote Tong, the president of Kiribati. A collection of islands sprinkled across the central Pacific, Kiribati is, for the most part, only a few feet above sea level, and the nation has already bought land in Fiji as an insurance policy against rising sea levels.

"For far too long, we have spoken of climate change as the most significant challenge," Tong said. "But what have we really achieved? What have we done about it?"

Before Figueres took her turn at the lectern, she carefully went through a printout of her remarks, crossing out the word *carbon* and replacing it with *emissions*. Again, she began by assuring the crowd—a mixture of climate negotiators, foreign ministers, and the occasional president—that a deal would be struck in Paris.

"In this moment, a climate change agreement is emerging," she said. Then she switched registers: "There is much political will already being displayed. But time is running out. And we must now turn that political will into clear leadership." She went on, "Ministers, this is your moment. Ministers here today, you and your peers, this is your moment."

The "bottom-up" approach has reduced the chances of an impasse, but it has not eliminated them. In the language of diplomacy, anything in brackets has yet to be agreed upon. The official negotiating text for the Paris COP is currently eighty-five pages long, and virtually everything in it remains in brackets, including the first word, "Preamble." Ten days of negotiations that were held in Bonn in June succeeded in paring the text down by just four pages; at this rate, to arrive at a treaty of, say, twenty pages would take months of nonstop talks. In his speech to the UN gathering, the secretary-general complained, "The key political issues are still on the table."

One of these issues is money. The hundred billion dollars a year that was promised in Copenhagen to poor countries is supposed to go partly toward helping them adapt to warming and partly toward financing climate-friendly energy systems. But almost everything about the financing remains unresolved, including where the cash will come from and what sorts of projects it will go toward. So far, wealthy nations have committed only about ten billion dollars to what's been dubbed the Green Climate Fund; this includes three billion dollars that the Obama administration has pledged but, because of congressional opposition, may not be able to make good on.

"There has to be a clear delivery of financial resources, because countries like us, we have very ambitious plans," Amjad Abdulla, the chief negotiator for a group of low-lying nations, the Alliance of Small Island States, told me by phone from the Maldives. "If we have to go back to communities and say, 'We couldn't get the money,' that's where they get furious."

258 / LIFE ON A LITTLE-KNOWN PLANET

"From a developing country's perspective, finance is going to be a deal-maker or a deal-breaker," Tosi Mpanu-Mpanu, a member of the negotiating team from the Democratic Republic of Congo, told me. "It is going to be the ultimate test of good faith."

Another issue is what's become known as "the gap." To hold warming to less than two degrees Celsius, global emissions would have to peak more or less immediately, then drop nearly to zero by the second half of the century. Alternatively, they could be allowed to grow for a decade or so longer, at which point they'd have to drop even more precipitately, along the sort of trajectory a person would follow falling off a cliff. In either case, it's likely that what are known as "negative emissions" would be needed. This means sucking CO_2 out of the air and storing it underground—something no one, at this point, knows how to do. The practical obstacles to realizing any of these scenarios have prompted some experts to observe that, for all intents and purposes, the two-degree limit has already been breached.

"The goal is effectively unachievable" is how David Victor, a professor at the University of California, San Diego, and Charles Kennel, a professor at the Scripps Institution of Oceanography, put it recently in the journal *Nature*.

Even those who, like Figueres, argue that the goal is still achievable acknowledge that the INDCs aren't nearly enough to achieve it. "I've already warned people in the press," she told the gathering at Citigroup. "If anyone comes to Paris and has a eureka moment—'Oh, my God, the INDCs do not take us to two degrees!'—I will chop the head off whoever publishes that. Because I've been saying this for a year and a half."

To deal with "the gap," many countries, as well as many of the groups that are unofficial participants in the negotiations, like Oxfam and the World Wildlife Fund, are pushing for what's become known as a "progression clause." This would commit countries to revising their INDCs every five years to make them more stringent. Among the countries that are opposing such a clause is Saudi Arabia.

"Players like the Saudis, like the Chinese right now, to be honest with you, are trying to water it down so you don't have a cycle of improvement," Jennifer Morgan, the global director of the climate program at the World Resources Institute, a research group, told me from her office in Berlin. "And that, I think, is the fight that's going to be the next three months. Do we get those kernels of integrity in the international agreement or not?"

ON FIGUERES'S LAST DAY in New York, I arranged to meet her at her hotel, not far from the UN. It was a purely functional place, with no lobby or bar, so we went up to an empty lounge on the top floor, where there was a microwave and a coffee machine. Figueres made herself a cup of black tea with a tea bag she'd brought from home. I'd brought along a list of questions on a piece of paper. A few minutes into our conversation, she took the paper from me and sketched out her vision of the future:

"I love this," she said. The straight line was supposed to represent economic growth, past and future, the curved line the rise and fall of greenhouse gas emissions.

"That's where we are," she said, drawing a dot right at the point where the two lines were about to diverge. She gestured toward an office tower across the street: "I think you and I will be alive when that building, all of those windows, will be covered with very, very thin-film solar cells, so that the building can produce all the energy it needs and maybe more."

I asked what would happen if the emissions line did not, in fact, start to head down soon. Tears welled up in her eyes and, for a moment, she couldn't speak.

"Ask all the islands," she said finally. "Ask Bangladesh. We just can't let that happen. Do we have the right to deprive people of their homes just because I want to own three SUVs? It just doesn't make any sense. And it's not how we think of ourselves. We don't think of ourselves as being egotistical, immoral individuals. And we're not. Fundamentally, we all have a morality bedrock. Every single human being has that."

Then she brightened: "You know, I think that this whole climate thing is a very interesting learning ground for humanity. I'm an anthropologist, so I look at the history of mankind. And where we are now is that we see that nations are interlinked, inextricably, and that what one does has an impact on the others. And I think this agreement in Paris is going to be the first time that nations come together in that realization. It's not going to be the last, because as we proceed into the twenty-first century there are going to be more and more challenges that need that planetary awareness. But this is the first, and it's actually very exciting. So I look at all of this and I go, This is so cool—to be alive right now!"

Originally published in *The New Yorker,*
August 24, 2015.

Update: Christiana Figueres now lives in Costa Rica. She is the cohost of the podcast Outrage & Optimism. *COP21 did produce a historic accord—the so-called Paris Agreement—but many countries, including the United States, have failed to live up to their Paris commitments.*

MR. GREEN

Environmentalism's Most Optimistic Guru

A MORY LOVINS'S HOME, which also serves as his office and "bioshelter," is open for self-guided tours weekdays from nine o'clock in the morning until four in the afternoon. Built into a mountainside above Snowmass, Colorado, it has curved stone walls, a flat roof, and several sets of solar panels, some of which rotate to track the angle of the sun. The building's double-paned windows are lined with a polyester film that allows visible light to pass in but prevents thermal radiation from getting out, and the space between the panes has been filled with krypton. Although wintertime temperatures on the mountain routinely drop below zero, the building has no furnace; it is warmed by sunlight and by heat that has been collected in, among other places, a pond that lies between the Xerox machine and the dining room. The first time I visited, Lovins had just finished doing some laundry in his front-loaded, energy-saving washing machine. He took the damp clothes out of the washer and hung them in a little glass-ceilinged room. It was a bright blue morning, and Lovins predicted that the clothes would be ready to wear by nightfall. In the winter, if the sky is overcast, it can take up to two days for items like blue jeans to dry completely, but this is no problem, he assured me, provided one is capable of thinking more than twenty-four hours in advance.

Lovins is a short man with a salt-and-pepper mustache, a fringe of tousled black hair, and droopy brown eyes that give him a passing resemblance to Einstein. He wears Coke-bottle eyeglasses, a necklace of turquoise beads, and a watch that is supposed to prevent jet lag by sending out an electromagnetic signal exactly the same frequency as the earth's. He is routinely described, even by people who don't particularly like or admire him, as a "genius."

Lovins first came to national attention in 1976, when he was twenty-eight. In an essay published in *Foreign Affairs,* he asserted that the United States could completely phase out its use of fossil fuels and do so not at a cost but at a profit. "We stand here confronted," he wrote, quoting Pogo, "by insurmountable opportunities." At the time, the country was in the midst of what might now be called the first energy crisis, and the article created a stir; testifying on Capitol Hill, Lovins emerged as the demand-side management version of a rock star. Symposia were held to debate his ideas, and critiques were published by, among others, the physicist and Nobel laureate Hans Bethe. (Lovins, in turn, wrote a response twice as long as Bethe's critique, and Bethe conceded several points.)

Thirty years later, the world faces another energy crisis, and Lovins still sees limitless opportunity. He maintains that the United States can eliminate its use of oil by 2050, even while reducing its coal and natural gas consumption, enjoying unprecedented prosperity, and preserving the Arctic National Wildlife Refuge. Although Lovins was one of the first to appreciate the dangers of global warming, he believes that the problem seems so daunting only because those studying it have got the math wrong. "Climate protection, like the Hubble space telescope, has been spoiled by a sign error," he told me.

Lovins is a prolific writer—of books, of articles, and of technical treatises. During my first visit with him, he informed me that he had picked out a few of the most important ones for me to take home: papers on topics like microgeneration, "super-efficient" building practices, and data-center design were arranged in stacks that covered nearly the entire surface of a large dining room table. That day, we ended up talking for several hours, and as I was packing up my things to go Lovins went to check on his laundry. It was nearly dry, he reported cheerfully. As I was driving back down the mountain in my rental car, it occurred to me that Lovins might be the most impractical person I had ever met. Then it occurred to me that he might be the only truly practical one.

THIS YEAR, Americans will consume close to four trillion kilowatt hours of electricity. In addition, we will burn through a hundred and forty-three billion gallons of gasoline, which at current retail prices will cost us some three hundred and sixty billion dollars, and twenty-six billion gallons of jet fuel, worth fifty billion dollars. To heat our homes and businesses this winter, we will purchase sixty-two billion dollars' worth of natural gas and heating oil, and just to grill our weenies we will buy some seven hundred and seventy-one million dollars' worth of charcoal briquettes. In 2007, total energy expenditures in the United States will come to more than a quadrillion dollars, or roughly a tenth of the country's gross domestic product.

With so much at stake, basic economics suggests that any significant inefficiencies should have been wrung out of the system long ago. It follows that further efforts will cost more than they will return. This reasoning is pervasive in the U.S., its most prominent spokesman being Vice President Dick Cheney, who once dismissed energy conservation as a "sign of personal virtue." Lovins's fundamental premise is that this fundamental premise is wrong.

"You know, there's this old joke about the economist who's taking his mannerly granddaughter for a walk," he told me.

"She says, 'Oh, Grandpa, I see a twenty-dollar bill lying in the street. May I go pick it up, please?' He says, 'Don't worry, my dear. If it were real, somebody would have picked it up already.'" Lovins likes to say that he takes economics "seriously, not literally." In his view, the streets are littered with twenty-dollar bills.

Lovins makes his living as the CEO of the Rocky Mountain Institute, a consulting firm that he founded twenty-five years ago with his wife at the time, Hunter. RMI used to operate out of Lovins's Snowmass home; in recent years it has outgrown these quarters—it now employs more than fifty people—and has expanded into new offices, some in Boulder and the rest down the road from his house, in a building that once belonged to a foundation created by John Denver. Lovins calls the firm an "entrepreneurial nonprofit," and its stated goal, which he often recites word for word, is to foster "the efficient and restorative use of resources to make the world secure, just, prosperous, and life-sustaining." Some of RMI's clients embrace this goal in its entirety, others at best selectively. (While I was visiting Lovins, he delivered the "efficient and restorative" spiel to a representative of the Singaporean government who had come to discuss manufacturing; her response was to giggle nervously.)

RMI, for its part, does not demand commonality of purpose. In "Why We Work with the Military," an essay posted on RMI's website, Lovins rejects the criticism—sometimes voiced by his own employees—that by consulting for the Department of Defense the institute is simply helping to kill people in a more energy-efficient manner. "A molecule of oil burned or carbon dioxide released has the same consequences no matter who used it," he observes. Other RMI clients have included San Diego Gas & Electric, Royal Dutch Shell, and Anglo American PLC, one of the world's largest mining companies. A few years ago, Texas Instruments hired RMI to help design a new chip-manufacturing plant in Richardson, Texas. It is expected to use 20 percent less energy and 35 percent less water than a typical chip factory of comparable size. It also cost 30 percent less to build.

"Amory doesn't take a bullying, negative approach," Paul Westbrook, Texas Instruments' manager for sustainable development, told me. "He just says, 'Here's a better way, and here's why it works.' And you think, Well, we'd be kind of dumb not to do that." One of the ways the new Texas Instruments plant will save energy is by capturing heat that normally would have been discarded as waste. "We implemented heat recovery, and, lo and behold, we didn't need as many boilers," Westbrook said.

A lot of Lovins's ideas sound radical and futuristic—ultralight cars made of carbon fibers, vehicles that generate electricity when they're not on the road, an economy powered by hydrogen. At the same time, he is a passionate advocate of what he calls "good, old-fashioned Victorian engineering," and he believes that a great many problems can be solved using high school physics. (Lovins can spend hours describing the energy savings that follow from steps as simple as increasing the diameter of pipes.) This combination of high- and low-tech enthusiasms makes his outlook difficult to categorize. Once, when I casually used the phrase "thinking outside the box," Lovins interrupted me. "There is no box," he said.

Perhaps RMI's most influential client these days is Walmart. Just to cart around goods that it sells in its stores, the company employs some sixty-eight hundred trucks, which annually consume at least a hundred and twenty-five million gallons of diesel fuel. In 2005, after consulting with Lovins, Walmart announced plans to double its fleet's fuel efficiency over the next ten years, from an average of six and a half miles per gallon to thirteen. Already, all of the company's trucks have been outfitted with auxiliary power units so that the driver doesn't have to keep the engine idling just to run the air conditioner.

"In a room of ten people talking about why it can't be done, Amory is the one working on the five ways to get there," Andy Ruben, Walmart's vice president for corporate strategy and sustainability, told me.

"I don't do problems" is how Lovins once put it to me. "I do solutions."

LOVINS, WHO IS FIFTY-NINE, grew up in towns along the Eastern Seaboard; when he was a child, his family moved from Silver Spring, Maryland, to Elmsford, New York, and then from Montclair, New Jersey, to Amherst, Massachusetts. His father, who designed optical equipment, spent a lot of time in his home workshop, tinkering, and in this way Lovins, too, became interested in gadgets. While he was still in high school, he built a nuclear magnetic-resonance spectrometer in his basement, and discovered what he calls a "peculiar and still unexplained solid-state effect" having to do with cobalt. When he went off to Harvard, he helped pay his way by doing consulting work in experimental physics for, among others, the Lincoln Laboratory at MIT. Lovins enjoyed college—in addition to physics, he studied law, linguistics, and chemistry. But when, in his junior year, he was told he would have to complete a major he dropped out and moved to England. He attended Oxford until he was once again pushed toward a prescribed course of study, at which point he quit school again.

By this time—1971—Lovins had come under the influence of David Brower, the charismatic founder of Friends of the Earth, and he went to work for the organization in London. One of Britain's largest mining companies, Rio Tinto, announced a plan to mine for copper in a national park in Wales, and at Brower's urging Lovins spent a year writing a book about the park. The book, *Eryri, the Mountains of Longing,* was instrumental in blocking Rio Tinto's plan. In the process of writing it, Lovins began to question the utility of his earlier research. (Today, Rio Tinto is a client of RMI's.)

"It gradually occurred to me that the problems I was working on, whether they were tertiary structure of proteins or straight physics, were interesting but not very important," he recalled. "Even understanding mitochondrial-membrane kinetics, which I briefly dabbled in, would be not nearly as important as solving basic problems of energy resources, environment, development, and security. Because it didn't much matter how well we understood these other matters if we weren't here."

Much of Lovins's early work centered on atomic energy. He wrote a series of papers arguing that the whole "atoms for peace" idea was misguided: there was no way to promote nuclear power without also promoting nuclear proliferation. (One of these papers spent two years under review by the U.S. government, which feared that Lovins had drawn the connection between processed fuel and bomb-making a little too clearly; the paper eventually appeared in *Nature.*) In his 1976 *Foreign Affairs* article, which was titled "Energy Strategy: The Road Not Taken?," Lovins urged that the U.S. stop exporting nuclear technology and, simultaneously, that it phase out its own atomic energy program. In the same piece, he warned that some of the alternatives to nuclear power were no less dangerous. At a time when the phrase *global warming* was barely in circulation, he observed:

The commitment to a long-term coal economy many times the scale of today's makes the doubling of atmospheric carbon dioxide concentration early in the next century virtually unavoidable, with the prospect then or soon thereafter of substantial and perhaps irreversible changes in global climate. Only the exact date of such changes is in question.

Lovins's opposition to both nuclear and coal-fired plants raised an obvious problem. How did he expect an energy-intensive economy like the United States' to function? The way out of this bind, Lovins argued, was to reimagine it. People weren't interested in energy for its own sake but, rather, for the benefits—hot showers, cold drinks, dry clothes—that it conferred. If Americans could get the same benefits using less energy, then they would, in effect, have found a new energy source. Meanwhile, instead of building large centralized power stations, they could gradually shift to localized sources of renewable power, like solar farms. Lovins labeled a future dominated by an ever-greater number of ever-larger power plants "the hard path"; the alternative he called "the soft path."

"The hard path entails serious environmental risks, many of which are poorly understood and some of which have probably not yet been thought of," he wrote. "The soft path . . . hedges our bets. Its environmental impacts are relatively small, tractable and reversible." Several years later, perusing a report put out by the Colorado Public Utilities Commission, Lovins came upon a misprint: someone had typed an *n* for an *m* in the word "megawatt." He coined another new term: *negawatt*. A negawatt is a watt of electricity that does not have to be generated because an energy-saving measure has obviated the need for it. By replacing a seventy-five-watt incandescent light bulb with a fourteen-watt compact fluorescent bulb, an individual can, for example, produce sixty-one negawatts. By replacing ten incandescent bulbs with ten compact fluorescents, the individual can generate six hundred and ten negawatts. Negawatts tend to produce more negawatts; for instance, a house lit with compact fluorescents requires less air-conditioning, since fluorescent bulbs emit a fraction of the heat of incandescents. The same principle can be applied to all forms of energy, including oil. Lovins likes to call the United States the "Saudi Arabia of nega-barrels."

• • •

THIS PAST FALL, Lovins came to Manhattan for a conference sponsored by former president Bill Clinton. The evening before the conference began, I went out to dinner with him at a Japanese restaurant in Midtown. We were ushered into a room in the back. A cone-shaped light fixture hanging from the ceiling cast a pallid gleam onto the table. A few feet away, a row of recessed lights threw circles of brightness onto nothing in particular.

"This is what happens if your lighting is designed by electricians," Lovins told me, glancing around. "Who wants spots of light on the carpet?" He noted that all the bulbs were incandescents and that the bulb hanging over the table was on a dimmer, which further reduced its already minimal efficiency. Lovins estimated that a better-designed system could cut the restaurant's lighting costs by 80 percent. "There's upwards of a hundred giant power plants to be saved by proper lighting systems," he observed, before turning his attention to the sushi menu.

To spend time with Lovins is to see the world as one long string of bad decisions. Waste and profligacy are everywhere: in inefficient lights, heat-leaking windows, gas-guzzling trucks, poorly designed eateries. It's not that people are stupid, exactly. It's that their intelligence is limited. When they make decisions, they tend to worry only about their own self-interest, which they see in such narrow terms that they miss the larger opportunities all around.

Take, for example, the electrical system of an average office building. "If we were to dig into the ceiling of most offices where the wiring is for the lighting, we'd probably find that the wire size was specced by the low-bid electrician to meet the National Electrical Code," Lovins told me. "The code says you need wire so fat for so much current. Well, it turns out that wire-size code is meant to prevent fires. What would be economically optimal in terms of resistance losses would be wire twice as fat, which means four times as much copper. Now, the electrician isn't going to pay your electric bills, right? If you had such an altruistic electrician that they were willing to put in four times as many pounds of copper to get you a one-year payback on your electric bills, they wouldn't get the job, because they wouldn't be the low-bid electrician anymore." The problem here is what's known as a split incentive but might better be called a mis-incentive. If the parties figured out how to divvy up the savings, they could both make money, but, because of ingrained habits, or a lack of creativity, these savings are never realized.

"Let me give you a specific case—a two-hundred-thousand-square-foot curtain-wall office tower near Chicago," Lovins said. "Chicago is cold in the winter and hot in the summer, and this was a very uncomfortable building all year round. In the winter, it had frost growing on the walls. The window seals were starting to fail, because they were twenty years old, so they were going to have to reglaze the whole glass curtain wall. Normally, you would put in the same glass that's already there, which in this case was dark, double-bronze, heat-absorbing glass with a gray film. It let in 9 percent of the light, so the place was as gloomy as a cave. We designed a super-window that would let in nearly six times as much visible light but a tenth less unwanted heat. It cost an extra seventy-eight cents per square foot of glass. If you combined those super-windows with retrofits that bounced the daylight all the way through the floor plate and with very efficient lights and lighting controls and office equipment, you could cut the peak cooling load on the hottest afternoon more than fourfold. That meant that instead of just renovating the big old air-conditioning system you could replace it with one that's four times smaller. And that would cost two hundred thousand bucks less, and that money could then pay for the better windows and the retrofitting of the lighting and other improvements. So you'd end up saving three-quarters of the energy at a slightly lower cost than the regular renovation—the payback time was minus five months. So we proposed that to the owner. They liked it, and we all thought it was going to be implemented. Surprise! It never happened. Why not? Because that particular building was controlled by a leasing agent whose incentive was deal flow. Whenever a floor got leased up, the agent would pocket a commission. And, not wanting to interrupt the commissions for a few months, the leasing agent vetoed the retrofit.

"It was a lovely lesson in how real such perverse incentives are and how pervasive," Lovins went on. "Each one is a showstopper, and each one is a business opportunity."

THE CLINTON GLOBAL Initiative is part of the Renaissance Weekend–Sun Valley–Davos circuit, and it combines, freely, elements from the Council on Foreign Relations and the Vanity Fair Oscar party. (On my way to meet Lovins, I happened upon Richard Branson salaaming, *Wayne's World* style, to Archbishop Desmond Tutu.) The day the conference began, Lovins slept in, missing speeches by Bill Clinton and Laura Bush, but showed up for a working lunch devoted to energy and climate change. He was wearing a black suit and carrying a duffel bag–size briefcase. At his table were an executive from Cisco, the head of a Norwegian philanthropy, and the chairman of a group called SmartTransportation.org, who told me that Lovins's work had inspired him to form the organization. The lunch was served on plates made from organically grown bamboo, and the table linens had been woven from compostable hemp. While the fair-trade coffee was being served, Warren Buffett walked in. Lovins unzipped his case, which turned out to contain, in addition to a laptop, a small library of RMI-produced books and papers. He picked out a three-hundred-page book titled *Winning the Oil Endgame* and made a beeline for Buffett. When Buffett declined to take the volume—he asked that it be sent to him at his office—Lovins seemed disappointed. But he recovered almost immediately. "That's smart of him," he whispered to me. "He doesn't take anything too heavy."

After lunch, everyone switched tables. This time, Lovins ended up sitting with several executives from Ford; Purnendu Chatterjee, the head of a private equity firm; and William McDonough, one of the preeminent "green" architects in the United States. In the front of the room, executives from the Swiss Reinsurance Company and DuPont were holding a panel discussion on energy-saving companies with the former NATO commander General Wesley Clark. Lovins spent most of the discussion sending emails. One was to the Norwegian philanthropist he had just met. The philanthropist owned a cheese farm;

Lovins attached a paper on energy-efficient dairy farming. When General Clark said something he disagreed with, Lovins sent him a long email outlining why. During a brief break in the program, Lovins sought out Thomas Friedman, the *New York Times* columnist, and Rick Fedrizzi, the president of the U.S. Green Building Council, to give them a PowerPoint on a school RMI had redesigned in Brazil. According to Lovins, the redesign had not only saved money but also improved students' test scores and, somewhat more mysteriously, their dental health. During another break, he presented the mayor of San Francisco, Gavin Newsom, with an inch-thick pile of articles. As Newsom took the articles, he laughingly alluded to an equally thick pile of reading material Lovins had handed him several weeks earlier, in Davos.

"Amory is my hero," José María Figueres, the former president of Costa Rica, told me. "He is just so much fun. I love the way he takes advantage of the opportunity to push his agenda in the most"—he paused for a moment—"wholesome way."

After the formal program ended, Lovins stayed on to talk to Purnendu Chatterjee. Among many other things, Chatterjee's firm holds a controlling interest in a refinery in India. Lovins told him that typically, in a refinery retrofit, RMI had been able to cut energy use by more than 40 percent. "Wow!" Chatterjee exclaimed. He said that he was thinking of building a new refinery. Lovins licked his lips and handed Chatterjee a stack of literature.

That night, there was a huge gala for conference participants at the Museum of Modern Art. Lovins arrived with another copy of *Winning the Oil Endgame* under his arm. It's not easy to eat a plateful of canapés while holding on to a three-hundred-page book, but somehow he managed. He chatted with the head of an Australian automotive company about vehicle efficiency; with an official of Habitat for Humanity about low-cost homes that can be made from bamboo; and with an executive from Coca-Cola (another RMI client). At one point, I told Lovins that a man who happened to be standing near us was one of New York's largest real estate owners. Lovins headed straight for him. He delivered his usual pitch about making money by saving energy. The man pointed out that as a landlord he didn't benefit from gains in efficiency, his tenants did. Lovins advised him to restructure his leasing agreements so that both parties could share in the gains. The man took Lovins's business card.

As the evening wore on, and most people fell to drinking and schmoozing with their friends, Lovins continued to work the room. When the dessert trays were being wheeled out, I lost track of him. By the time I caught up with him again, he had managed to give away his book.

WINNING THE OIL ENDGAME is a characteristic Lovinsian project—ostensibly hard-nosed and at the same time shamelessly utopian. The book's foreword is by former secretary of state George Shultz, who notes that "Amory Lovins loves to be a bull in a china closet—anybody's china closet," and the epigraph is by Antoine de Saint-Exupéry: "If you want to build a ship, don't drum up the men to gather wood, divide the work and give orders. Instead, teach them to yearn for the vast and endless sea." The book's premise is a variation on the theme Lovins first laid out back in 1976, this time applied to petroleum. The United States, he argues, can cut its oil imports to zero by 2040, eliminate oil use entirely by 2050, and make money in the process.

In December, Lovins traveled to Washington, D.C., to give a talk on the book. I met him at the airport, and on the way into town we stopped at his hotel to drop off his suitcase. When Lovins opened the door to his room, he was met by a blast of hot air. The room had its own thermostat, which had been set to seventy-nine degrees. A lamp by the desk was burning cheerfully. Lovins peered under the shade. "It's an incandescent," he said, switching it off.

Lovins's speech took place in a hotel not far from the Pentagon. The crowd that greeted him in the ballroom was an eclectic mix: representatives from organizations like the Northern Virginia Conservation Trust and the Post Carbon Institute were sitting next to officials from the Defense Department and the Government Accountability Office. (The research for *Winning the Oil Endgame* was partly funded by the Office of the Secretary of Defense.)

Lovins began with a slide showing an army tent on a patch of sand—presumably in Iraq or Afghanistan. The tent was completely uninsulated. In front of it stood a five-ton air-conditioning unit, and somewhere in the distance an oil-fired generator was producing the electricity to power the unit. The oil for the generator had had to be hauled a great distance, at great peril to those guarding it, and yet most of the energy was going, in Lovins's words, into "air-conditioning the desert."

From the desert, Lovins turned to the seas to tell one of his favorite stories, about whale hunting. "In 1850, the fifth-biggest industry in our country was whaling, and most houses were lit by whale-oil lamps," he said. "But as whales started to get shy or scarce and the price of whale oil drifted up, this started to elicit competition, particularly from coal-based oil and gas." By 1859, these competitors had seized five-sixths of the whale oil–lighting market. "This was a real shock to the whalers. They never expected to run out of customers before they ran out of whales. But that's what happened, and they were soon reduced to begging for subsidies on national security grounds.

"Oil feels a little like this now," Lovins continued. "We've spent over thirty years amassing a very powerful portfolio of new ways to save oil or substitute for oil, but no one had bothered to add it up. And when we did so we found it was enough to save all the oil we use and more."

The largest share of oil consumed in the U.S.—nearly 70 percent—is eaten up by transportation. By Lovins's reckoning, most of that is wasted. The power needed to propel a car is a function of its weight, so building a vehicle out of steel means using energy largely to move metal. Lovins proposes making cars instead out of ultralight carbon composites. An ultralight car with a Prius-style hybrid engine could—in theory, at least—get upward of seventy miles per gallon. Lovins has been pushing an ultralight vehicle of his own invention, which he calls a Hypercar, for the past fifteen years, and has created a company, named Fiber-forge, to license the technology needed to manufacture the necessary parts. But so far no one has been willing to finance a prototype. (Lovins himself gets around in a hybrid Honda Insight, which sports the bumper sticker "I'd rather be driving a Hypercar.")

According to Lovins, by switching to ultralight vehicles (including airplanes) and implementing a variety of other "end-use efficiency" technologies the U.S. could eliminate half of its oil needs. It could eliminate another 20 percent by substituting biofuels for oil, and the last 30 percent by replacing oil with natural gas. (Saving enough natural gas to replace a third of the country's oil could be easily accomplished, he maintains, by, among other things, reducing electricity consumption.) The cost of eliminating oil use entirely would, by his calculations, come to half of what, by official forecasts, would be spent on purchasing it. Meanwhile, the U.S.'s CO_2 emissions would drop by 25 percent.

"We know this sort of thing works, because we've done it before," Lovins assured the crowd in Washington. He gestured toward a graph of oil consumption that had been projected onto a screen behind him. "Look at how steeply oil use and imports fell the last time we paid attention, which was from 1977 to 1985. In those eight years, the economy grew 27 percent while oil use fell 17 percent.

"All this can be done with all the same economic growth and the same growth in driving, in flying, in huge houses, and so on, that's in the government forecasts," he continued. "No new invention and no changes in lifestyle. And it could all be done without new taxes, mandates, subsidies, federal laws, or anything else either party doesn't like or could mess up." Lovins went on in this vein for nearly ninety minutes, ending with a quote from Marshall McLuhan: "Only puny secrets need protection. Big discoveries are protected by public incredulity."

Lovins's talk had gone on so long that there wasn't much time for Q & A. The few questions asked—most of them comments, really—ranged from mildly skeptical to aggressively so. One man demanded to know why Lovins would not "accept the fact" that there had been many advances in nuclear power safety. A second man stood up and said that as long as he could afford it he was going to continue to drive his Jeep, his Jaguar, and his Harley. He said that the only way he could see to move the United States off oil was to raise the price. Afterward, while Lovins was handing out more copies of *Winning the Oil Endgame,* I got to talking to David Goldstein, the president of the Electric

Vehicle Association of Greater Washington. He said that he did not expect to see ultralight cars on the road anytime soon.

"Amory doesn't understand some of the practical realities of how the world works," he told me. "I want to believe him, but . . ."

Lovins's promise that apparently intractable problems—oil dependence, global warming, nuclear proliferation—can be profitably resolved is both the great appeal of his approach and its biggest liability. Much of what he recommends sounds just too good to be true, the econometric version of "Shed pounds by eating chocolate!"

John Holdren, the president of the American Association for the Advancement of Science, has known Lovins for more than thirty years and considers him a friend. He says that Lovins's contention that a great deal more work could be wrung from the energy we consume is "absolutely unassailable."

"Amory has been more energetic, more persistent, and more creative in thinking about ways to make this happen than anyone else," he told me. "But Amory has always believed—seemed to believe—that free-market economics alone, rationally applied, will solve the problem. And I don't believe that." Holdren observed that current projections suggest that global energy consumption will reach eighteen hundred exajoules by 2100. In order to hold atmospheric CO_2 levels at a "tolerable" level, the world will need to produce five times the amount of energy it now gets from fossil fuels, using sources that produce no carbon dioxide. The only conceivable way to achieve this, he said, is to make the use of fossil fuels much more costly. "We just aren't going to do it unless there's a price on carbon."

Marty Hoffert, a professor emeritus of physics at New York University who has written extensively on energy demand and global warming, is also on cordial terms with Lovins. "I support virtually all of Amory's calls for efficiency," Hoffert told me. "It's an important thing to have people thinking about efficiency. But I don't think that's going to get us there. If we do things more efficiently, people may just consume more."

This hypothetical problem is not merely hypothetical. Since the mid-nineteen-seventies, the American economy has grown steadily more energy efficient. What's called the economy's "energy intensity"—the ratio of BTUs to GDP—has been dropping by roughly 1.5 percent a year, for a total of nearly 46 percent over three decades. These savings have, however, been more than offset by people thinking up new ways to use energy, so that in that same period the country's total energy consumption has risen by 39 percent and its CO_2 emissions by roughly the same proportion.

Meanwhile, the example Lovins likes to point to—the drop in oil use in the early nineteen-eighties—is, at best, equivocal. As Lovins notes in his book, what made people "pay attention" to oil consumption was the 1973 Arab oil embargo and the second, even more severe 1979 oil shock. Part of the drop was due to structural shifts in the economy away from oil-intensive activities. Part of it was fuel substitution by both individual consumers and industry, as homeowners and factories switched from oil to, for example, natural gas. The largest part of it was increased fuel efficiency in both automobiles and buildings, led by the creation of federal auto efficiency standards in 1975. And, finally, part of it was a change in consumer behavior as Americans bought smaller cars and turned down their thermostats. Thus what Lovins offers as a demonstration that federal regulation and new taxes are unnecessary could just as plausibly be seen as evidence of exactly the opposite.

Lovins knows all these arguments and is unmoved by them. "Sometimes after I give a talk, some folks get irked that I talk only about solutions and not about problems," he told me. "And typically someone will get up and give a long riff about all the bad things happening and all the suffering in the universe, which is basically true. And the only way I've found to deal with that sincere and well-founded concern is to let the person run down after a while and then ask as gently as I can whether feeling that way makes them more effective.

"What I think people are most hungry for—and trying hardest to locate—is practical and profitable solutions. And that's where I can make a distinctive contribution. It's a very pragmatic approach."

After his speech, Lovins still had, by his accounting, eight hours of work remaining on a report that was due the following day. I went with him to his hotel to ask a few last questions. This time, when he opened the door the room was freezing; by turning down the heat he had apparently turned on the air conditioner. I asked Lovins how his plan to save the world through energy efficiency could accommodate the open-ended nature of human desire. If, as he claims, conservation is profitable, what was to stop the profits from going straight toward more consumption?

"It doesn't automatically prevent that," he said. But, he added, "you might plow the money back into more efficiency rather than more powerboats and helicopter skiing. After all, you don't rewash your clean clothes in the cheaper-to-run washing machine, because your clothes are already clean. At some point, I think you get jaded by continuous trips to Bali.

"Your neighbors might point out that what you're doing is increasingly antisocial," he continued. "On a moral or spiritual level, at some point you may discover you're not all that happy having more stuff or more travel. Trying to meet nonmaterial needs by material means is stupid and futile. Every faith tradition that I know decries materialism.

"Markets are meant to be greedy, not fair. Efficient, not sufficient. They're very good at short-term allocation of scarce resources, but that's all they're good at. They were never meant to tell you how much is enough or how to fulfill the higher purpose of a human being."

It was getting late, and I was keeping Lovins from his report. As I prepared to leave, he recalled another line, this one from Wallace Stevens: "After the final no there comes a yes and on that yes the future world depends."

Originally published in *The New Yorker,*
January 22, 2007.

Update: Amory Lovins is now the chairman emeritus of the Rocky Mountain Institute, which has grown to employ some seven hundred people on four continents.

THE GURU OF DOO-DOO

Profile of Sam Wasser, Who Uses DNA
to Fight Elephant Poaching

WHEN SAM WASSER, a conservation biologist at the University of Washington, flew into Singapore on October 26, 2015, one of the first things he did was borrow a car and drive to a hardware store. He bought two circular saws, some F-clamps, and a wheelbarrow. Then he headed to an old aviary inside the city's port.

Authorities, acting on an anonymous tip, had seized eighty tea sacks that had been shipped from Mombasa, Kenya. The tea had been replaced with elephant tusks—seventeen hundred of them, which together weighed more than five tons. On his first day at the site, Wasser measured and weighed each tusk individually. The weather was hot—almost ninety degrees—and muggy, and within a half hour Wasser was drenched in sweat. He logged the weights on an Excel spreadsheet and recorded any unique markings on the tusks. (Several had large black x's, which presumably identified them as belonging to a certain dealer or poacher.) The next day, his team tried to match pairs of tusks—he didn't want to sample the same elephant twice—and clustered them into groups. Finally, he put on a mask and goggles, unboxed the circular saws, and started cutting out pieces of ivory about the size of a matchbox.

At one point, he gestured at the hundreds of tusks—beige and brown and rust-colored—laid out on the ground. His T-shirt was dripping wet and covered in a fine layer of ivory powder. "There are periods throughout the process where it really gets to me," he said.

"Especially that there is such a large number of tusks weighing less than one pound—too small even to sell. I mean, look at all this. It's insane." He estimated that the seizure represented at least a thousand dead elephants.

Wasser, who is sixty-three, has black hair, a graying beard, and, under his eyes, the deep, dark circles of the chronically sleep-deprived. Two decades ago, he began working on a geographic map of elephant genetics using DNA extracted from dung. Populations from different regions carry different mutations, and Wasser's map shows where each mutation can be found. When he analyzes a piece of ivory, he can identify its specific mutation and match it up with his dung map, locating the spot where the animal was slaughtered. It's like having a high-level informant inside the world of wildlife crime.

Over the past several years, as elephant poaching has reached crisis proportions, Wasser has found himself ever more in demand. His work on elephants is financed, in part, by the family foundation of Microsoft cofounder Paul G. Allen. (Wasser has also received funding from the Smithsonian Institution, as well as from the U.S. Department of State and the UN Office on Drugs and Crime.) His sampling efforts have shown that most illegal ivory is coming from just two "hot spots." This, in turn, has provided a powerful tool for law enforcement officials deciding where to focus their resources. And it has helped lead to some high-level arrests, including one of an ivory dealer from Togo nicknamed the Boss and a Chinese woman who'd been dubbed the Queen of Ivory.

"I can't say enough good about his research and what he's done," Susan Snyder, director of the Office of Anti-Crime Programs at the State Department, told me.

"I think Sam deserves a Nobel Prize," said Bill Clark, a former chairman of Interpol's Wildlife Crimes Group.

"THIS IS KILLER WHALE," Wasser said, pulling a large plastic test tube out of a freezer. He was back in his lab in Seattle, though only briefly, because he was about to head off to Geneva for a convention on illegal wildlife trafficking. At the bottom of the tube sat half an inch of what looked like frozen mud but was, in fact, orca ordure. It had been collected with the aid of a specially trained dog named Tucker, who can sniff out floating whale droppings from a mile away.

"We've got Hawaiian monk seal in here," Wasser went on, indicating another tube. "Polar bear. Pacific pocket mouse. Sage grouse."

"Here's caribou," he said. He proffered a visitor a baggie filled with brown pellets.

Wasser has been called the "guru of doo-doo," and it's a title he wears with pride. In the nineteen-eighties, he pioneered the use of scat as a tool for studying wild animals by extracting hormones from their droppings. Then, in the nineteen-nineties, he became one of the first researchers to show that feces could be a source of DNA. "Scat is the most accessible animal product in the world," Wasser told me. "And it contains a huge amount of information, from the DNA of the animal that left it, to the DNA of all the things the animal was eating, to the microbiome in its gut, to its reproductive hormones, to its stress and nutritional hormones, to toxins."

Because scat contains so much information—and because so much is churned out daily—Wasser has been able to resolve questions that probably otherwise would have been unanswerable. When, for example, orcas off the San Juan Islands stopped having babies, no one was sure why. Some marine biologists blamed stress caused by boatfuls of whale-watching tourists; others proposed the cause was toxins, like PCBs, which accumulate up the food chain. By analyzing the orca poop from the open-bowed deck of a Grady-White powerboat, Wasser was able to determine that the orca whales were conceiving. The trouble was they were miscarrying 60 percent of their fetuses. The reason: a decline in the whales' favorite food, Chinook salmon. As the orcas grew hungrier, their fat released toxins that ended their pregnancies. (The discovery has not yet led to any policy changes, to Wasser's regret.)

Wasser began enlisting dogs in his research after he attended a conference on bears and heard a talk about hunting with hounds. He found a program run in a state prison that taught dogs how to sniff out narcotics, and the sergeant in charge invited him to attend two rounds of training. "They start the dogs on marijuana because it smells so much," Wasser explained. "When they moved to heroin, we moved to poop."

In one study, Wasser used scat-sniffing dogs to track five large animals—giant armadillos, giant anteaters, maned wolves, pumas, and jaguars—through Brazil's Cerrado, a once-vast savanna that's largely been cut up into ranches. The study found that jaguars and giant armadillos were particularly sensitive to human disturbance and avoided agricultural land. Maned wolves, pumas, and anteaters, meanwhile, were attracted to woodland and forest vegetation remnants found within ranch lands.

He and his grad students are now working on a DNA map of pangolin poop. The only mammal wholly covered in scales, pangolins look like a cross between a badger and an artichoke. As many as one hundred thousand of them are poached each year, making them the world's most heavily trafficked mammal. In parts of Asia, pangolin scales, which are made of keratin, like your fingernails, are prized for their medicinal value (though, according to science, they don't have any). Pangolin is also regarded as a delicacy, particularly in high-end restaurants in Vietnam, where the meat can sell for more than one hundred and fifty dollars a pound. To make his pangolin map, Wasser is planning to send specially trained dogs to Southeast Asia to sniff out droppings.

Wasser's lab currently holds several freezers filled to bursting with animal feces. Next to one of them I noticed a six-inch-high figurine of Tommy Chong—of Cheech and Chong comic fame—whom Wasser, broadly speaking, resembles. Someone had outfitted the figurine with a tiny T-shirt that said "I ♥ Scat."

WASSER GREW UP in Detroit, and from early on, he knew what he wanted to do. "I was determined to be a wildlife vet in Africa," he said. One summer in college, he got a job with a researcher who was studying lions in Uganda. En route, Wasser stopped in Nairobi, Kenya. There he found a letter waiting for him. It said dictator Idi Amin's soldiers had raided the researcher's camp, stolen his truck, and destroyed his data. "Don't come," it instructed. Wasser hadn't traveled halfway around the world just to turn around and go home, so he found a job on a different lion study, in Kenya.

Eventually Wasser found himself working with baboons in Tanzania. Watching them, he became convinced that dominant females were forming aggressive coalitions, preventing other adult females of the troop from becoming pregnant. He theorized this was to improve their own offspring's chances of survival. But it was hard to confirm his idea. This difficulty led him to the work of a cancer researcher who was tracking his patients' hormone levels by analyzing stool samples. It was Wasser's introduction to the power of poop.

Baboons range widely, and following them Wasser covered a lot of territory. On his travels, he began to encounter gruesome relics—sometimes an elephant skull, sometimes a whole carcass. One find in particular stuck with him: a pair of teeth—the first very small and the second enormous. Probably poachers had shot a baby elephant, waited for its mother to come defend it, and then shot her. "That was a turning point for me," Wasser said. "I was on a mission."

Elephant ordure, Wasser knew, wasn't hard to come by. "Often when I was working in the field, I'd just pull up a dried elephant poop and use it as a chair," he recalled. "It was everywhere." Meanwhile, a single gram of scat might contain millions of sloughed-off cells, each with a copy of its producer's DNA. Wasser started to collect samples on his own. Then he put out a call to biologists and game wardens all around Africa: send me your elephant scat. In this way, the groundwork was laid for his map.

Every elephant's DNA is similar to every other elephant's, just as my DNA, or yours, is similar to that of every other person on earth. But different elephant groups carry different mutations. These mutations tend to build up in non-protein-coding parts of the genome—so-called junk DNA. These are the segments Wasser focused on. He located sixteen stretches of elephant DNA where animals from different regions carry different numbers of repeating segments. (The stretches are known as microsatellites.) After ten years and thousands of analyses, Wasser reached the point where, presented with a blind sample of elephant dung, he could tell where it had been collected, within a hundred and ninety miles.

The first chance Wasser got to put his map to use came in 2005. Authorities in Singapore had seized a shipment of more than seven tons of ivory. The shipment, marked "soapstone," had traveled by ship from Malawi to South Africa, and from there had been transferred to a boat bound for Asia. It contained more than five hundred whole tusks and some forty thousand small ivory cylinders. (The cylinders were clearly intended for use as hanko signature stamps, which are popular in Japan and China.) Among law enforcement officials, the assumption was that to put together a shipment this large, a dealer had to have spent years amassing ivory from many different regions. But the DNA analysis proved otherwise. All the tusks could be traced to a single population of elephants concentrated in Zambia.

"We showed the poachers were going to the same place, over and over again, and that it was likely the ivory was relatively new," Wasser said.

The following year, in Taiwan, officials grew suspicious about two shipping containers. The containers, ostensibly filled with sisal, were supposed to be headed to the Philippines, but they seemed to keep bouncing around Asia; on the same voyage, they'd already passed through Taiwan once before. When the customs officials pried open the containers they found eleven hundred elephant tusks. Just a few days later, in Hong Kong, a resident's complaint about a stench coming from a neighboring warehouse led to the discovery of another four hundred tusks. Wasser analyzed the contents of both seizures. Again, he showed the ivory had all come from the same region, in both cases from southern Tanzania. A pattern was starting to emerge.

IN THE MID-NINETEEN-SEVENTIES, when Wasser first started working in Africa, roughly one and a half million elephants roamed the continent. Over the next decade and a half, the value of ivory, which at that point still could be legally traded, skyrocketed. During the nineteen-eighties, the price more than quintupled, from about twenty-five to one hundred and thirty-five dollars a pound. The elephant population, meanwhile, plummeted; by 1989, it had fallen to around six hundred thousand, and experts warned that Africa's elephants were headed toward extinction.

To reverse this gruesome trend, parties to the Convention on International Trade in Endangered Species of Wild Fauna and Flora, or CITES, enacted what amounted to a ban on international sales of African ivory. The ban went into effect in 1990, and for several years it seemed to be working. Poaching eased, and in some parts of Africa, elephant populations started to recover. But in 2006, just after Wasser began putting his map to use, the killing started up again. Growing demand in Asia lofted prices to new levels. By 2012, black market ivory was fetching one thousand dollars a pound in Beijing. That year alone, an estimated twenty-two thousand African elephants were poached. Clearly deaths were outpacing births, and, once again, experts warned of a crisis.

"The question is: Do you want your children to grow up in a world without elephants?" is how Andrew Dobson, an ecologist at Princeton, put it. The number of elephants in Africa may now be about four hundred thousand, which means that if current trends continue, the animals could be wiped out within two decades.

For Wasser, the new wave of poaching translated into a flood of samples. The seizures in Taiwan and Hong Kong were followed by seizures in, among other places, the Philippines, Thailand, and Malaysia. By 2015, he'd analyzed twenty-eight major shipments, totaling more than sixty-two and a half tons of ivory. The results were depressingly consistent. Each shipment had a clear geographical signature, which indicated that all, or at least most, of the tusks had been amassed from a single region. And the same signatures kept showing up over and over. The vast bulk of the ivory came from elephants in two regions. The first is an area known as the Tridom, which includes parts of northeastern Gabon, the northwestern Republic of the Congo, and southeastern Cameroon. The second region includes parts of Tanzania, primarily the area where Wasser used to study baboons, as well as parts of northern Mozambique and southern Kenya.

Right around the time I visited him in Seattle, the Great Elephant Census released its preliminary results. Researchers had conducted aerial surveys across Africa, collectively flying some two hundred and eighty-eight thousand miles. The census lined up with Wasser's findings: between 2009 and 2016 the number of elephants in Tanzania fell by more than half, from one hundred and nine thousand and fifty-one to forty-two thousand eight hundred and seventy-one.

"I kept expecting it to change," Wasser told me. "About a year and a half ago, I thought, Oh my God, I'm looking at all the seizures, and every single one of them is coming from the same place."

Two years ago, Wasser and his colleagues noticed that more than half the tusks in a given seizure were loners—the other tusk from that animal wasn't in the same shipment. Using DNA analysis, they were able to find the match for these tusks, often sent several months earlier or later, but always going through the same port. This indicated that two shipments had been packed by the same trafficker. "In doing that, we've been able to find major trafficking networks and track their sizes," Wasser said.

A picture is now starting to emerge. Major dealers or their middlemen supply poachers with weapons and purchase orders: send us this much ivory by this date. Poachers hunt in a concentrated area, filling the order bit by bit: two tusks on a motorcycle, ten in a car, until the quota is met. The kingpins sit removed from it all. They often try to outsmart customs officials by shipping the ivory from a neighboring country and moving it through four or five ports before it reaches its final destination. Wildlife divisions on the ground have confirmed many of Wasser's hypotheses. "You know you're close to the right answer when you're talking to officials on the ground and they say, 'That makes sense. We've seen this and this is going on, and it fits perfectly with what you're saying.'"

AFTER WASSER IDENTIFIED Tanzania as Africa's largest poaching hot spot, he grew nervous about returning to the country. His fears were heightened when an official he'd worked with closely there was murdered.

In November 2015, though, things started to change. Tanzania swore in a new president, John Magufuli, who started making serious efforts to improve intelligence and crack down on traffickers. Wasser returned to Tanzania this year as part of a training program conducted with the World Customs Organization. More recently, Tanzania gave Wasser and his colleagues permission to sample the three largest shipments they'd seized. "To me, that is a monumental step forward," he said. "They're telling us, 'We're ready to work with you to solve this problem.' A lot of this, I have to believe, is because of all the international attention that's been put on Tanzania as a result of our findings."

Now Wasser's main concern is that the world's ivory hot spot will relocate. The process will take time: traders will need to find a new country with sufficient elephants, learn where the elephants congregate, and set up new distribution chains.

The way to squelch this process, Wasser says, is for governments to provide samples from their ivory seizures, and to do it far more quickly. "Most countries don't turn their shipments over to us until a year, sometimes two years, after they've been seized," he said. "If they gave us more recent seizures, we could identify emerging hotspots. If you wait till these places get entrenched in corruption, the network becomes far more difficult to dismantle."

He's hopeful that more countries will start cooperating. In November, Vietnam decided to destroy a two-ton seizure in its stockpile and let Wasser sample it. "This was a huge breakthrough," he said. "Vietnam has seized thirty tons of ivory since 2010, and this was the first time they've done anything like this," he said. He also trained local officials how to do the sampling themselves. "The more countries start seeing the kind of information this is providing, the more willing they are to provide samples from their seizures," he said. "This battle is so hard to win. But it feels like we're on the verge of making a really big difference."

Originally published in *Smithsonian Magazine* as "The Elephant Detective," January–February 2017.

LAST WORDS

The Race to Save Eyak,
a Dying Language

THE LAST FULL-BLOODED member of the Eyak nation, Chief Marie Smith Jones, lives in Anchorage, Alaska, in a two-story, tan-colored building posted with a notice that warns "No dogs" and "No waterbeds." On a gray Saturday last spring, I arranged to go visit her. When I arrived at her apartment, no one answered the door. I thought perhaps she had forgotten our appointment and had gone off, but when I called up from the street it turned out that she had decided she did not feel like talking to me that afternoon, or, she implied, ever. I probably never would have got to meet her except that a friend of hers had advised me to bring along, as a gesture of respect, some halibut, a fish prized by the Eyak. Smith Jones asked me if the halibut could fit inside the mailbox. I said I didn't think so.

The Eyak are a mysterious people. By the eighteenth century, they were living near Prince William Sound, in close proximity to other, more formidable nations. (The Chugach called them Ungalarmiut, meaning "the people living to your left as you face the ocean," or, in this case, "Easterners," and the Ahtnas called them Danggane, meaning "uplanders.") Yet for millennia they somehow managed to maintain not only their own culture but also their own language: Eyak is as closely related to Navajo as it is to any tongue spoken by Native Alaskans today. In addition to being the last full-blooded Eyak, Smith Jones is the last remaining Native Eyak speaker. When I asked her how she felt about this, she said, "How would you feel if your baby died? If someone asked you, 'What was it like to see it lying in the cradle?' So think about that before you ask that kind of a question."

Smith Jones is eighty-seven and legally blind, with white, wispy hair, a face covered in fine wrinkles, and wrists no wider than a child's. Our meeting that afternoon ended up lasting for several hours and consisted in large measure of exchanges like this. At one point, Smith Jones told me that, like most white people, I had an "iron face"; at another, she announced, "I'm sorry to say, but I hate reporters." In response to a question about an appearance she had made a few years ago at a United Nations conference on Indigenous peoples, she brought up the Dalai Lama: "Beautiful words come out of his mouth, but his actions tell differently." In between such pronouncements, Smith Jones periodically softened. She talked about being sent off to work at a cannery at the age of twelve, about her struggles with alcoholism, about the stories her mother used to tell her, and about the difference in taste between a fish that has been caught in a net and a fish that has been harpooned, so that the blood slowly drains from its body. Smith Jones is a heavy smoker, and whenever there was a lull in the conversation she would light up another Winston and start to cough—a deep, hacking, raspy cough. Arrayed around her in her living room were at least a dozen prescription drug bottles and an unopened Sierra Club calendar from 1992. As I was getting ready to leave, she pulled a small card out of her wallet. On it was written her name in Eyak—Udach'Kugaxa'a ch'—which, she said, translated as "a sound that calls people from afar." I asked her again about being the last Eyak speaker, and this time she responded differently.

"I got that strong feeling right here that it's going to come back," she told me, putting a hand over her heart. "God will send down Eyak to start all over again."

LANGUAGE IS OFTEN described as man's essential accomplishment, yet nothing threatens the world's languages so much as human progress. A great wave of language extinctions, it is believed, took place eight or nine thousand years ago, following the so-called Neolithic revolution. As people gave up foraging in favor of farming, they started to live in larger and less isolated communities; certain groups prospered and became dominant, while others—and their languages—were absorbed. A second great wave of extinctions began with Europe's colonization of the Americas and Oceania. It is estimated, for example, that before the arrival of the first white settlers in Australia, in 1788, some two hundred and sixty aboriginal languages were in use. Traces of only a hundred remain. By the end of this century, it is likely that more than half of the six thousand languages still spoken around the globe will have vanished.

Besides Marie Smith Jones, the only other person who understands Eyak is a linguist named Michael Krauss. Krauss, who is seventy, lives in Fairbanks, but when I phoned him last winter he had recently been given a diagnosis of cancer and was preparing to go to New York to consult with doctors about a course of treatment. "What's happened is very ironic," he told me. "I made a proposal to the National Science Foundation to study Eyak in 1963. In it, I remarked there were only a few speakers left and the youngest"—Smith Jones—"was already, I think I said, 'elderly.' Now this 'elderly' lady calls me in alarm over my health."

I finally met up with Krauss in Seattle, a few months after we first spoke. He was staying with his daughter, and convalescing after surgery. Krauss has an open face, wavy white hair, and a manner at once passionate and professorial. Until his illness, he had had a broad, even stout, figure; now he was painfully thin. His daughter kept urging him to eat something. He kept saying that he would, and then leaving the food that she brought him untouched.

Krauss traces his interest in linguistics back to his childhood. Growing up in Cleveland, he was sent off to Hebrew school, where he studied a language that had nearly died and then been resurrected. After Hebrew faded as a spoken language, around two thousand years ago, it was preserved ceremonially and transmitted through texts. Then, in the late nineteenth century, the language was revived by early Zionists. So successful was this revival that today it is the preferred tongue of at least three million people. "It's a very remarkable history of survival against odds," Krauss told me. "It inculcated in me a real sense of what language means to nationhood and the survival of identity." Later, Krauss did graduate work at Harvard and learned, among other languages, Gaelic, Icelandic, and Faroese, the last of which is spoken only by the fifty thousand or so inhabitants of the Faroe Islands, northwest of Scotland. He became fascinated by the way that some languages, like Faroese, continued to thrive even as many other, much more broadly based languages were vanishing. In 1960, Krauss moved across the country to start the Linguistics Department at the University of Alaska Fairbanks.

As soon as Krauss got to Fairbanks, he set about assessing the condition of the state's Native languages. For nearly three-quarters of a century, children in Alaska had been prohibited from studying, or even speaking, their own languages at school, under a policy established by the territory's earliest commissioner of education, Sheldon Jackson. "Pupils are required to speak and write English exclusively," Jackson, who was also a Presbyterian missionary, wrote in 1888. "Instruction in their vernacular is not only of no use to them but is detrimental to their speedy education and civilization." As a result of this policy, which endured until 1972, virtually all the Native languages in Alaska fell into decline. Most imperiled of all was Eyak. Krauss was immediately drawn to it, both because of its fragility and because of its tenacity.

"We must not forget that all the languages we find today, even moribund relics like Eyak, are themselves the relatively few victors in the constant struggle for linguistic survival," he later wrote, for they have lasted "long enough to be at least documented, as compared with the many more languages which have disappeared forever without a recognizable trace."

EYAK IS NEITHER a linguistic isolate, like Basque, nor a part of any modern language community. Rather, it might be described as the spinster aunt of the Athabascan language group, whose members include Navajo and Apache. All Athabascan languages descend from a common ancestor—now extinct—that linguists call proto-Athabascan, which is believed to have originated in the interior of Alaska or the

Yukon Territory around the time of Christ. Eyak is thought to have broken off from the even more ancient linguistic branch that led to proto-Athabascan sometime around 1000 B.C. (A technique known as "glottochronology"—the phonetic analogue of carbon dating—can be used to determine the point at which related languages diverged.) For Eyak to have maintained its distinctive character for so long, its speakers must have spent the better part of the past three thousand years living essentially in isolation. One of the many unanswered questions about them is where.

Various clues within the language itself suggest that the Eyak were, at one point, an inland people. Their word for "downriver," for instance, *li'*, also means "into the closed end of something," hinting that the Eyak's route to the sea was blocked, perhaps by glaciers. By the time the first whites arrived in Alaska, though, the Eyak were living along the southern coast, from Prince William Sound down to what is now the town of Yakutat. The Russians, who were trying to establish a network of trading posts in the area, encountered them in the seventeen-nineties. Relations between the two peoples were occasionally tense; in 1799 a group of Eyak massacred a Russian hunting party, in retaliation for which the Russians tortured and killed an Eyak man. Nevertheless, some Eyak were converted to the Russian Orthodox Church.

Although few reliable records from the nineteenth century survive—if they ever existed—Eyak territory seems to have shrunk dramatically during that period. Several villages disappeared, probably wiped out by smallpox or measles. Others were assimilated by the Tlingit, a friendly but more powerful tribe that was expanding from the southeast. (If the English-speaking Americans hadn't appeared, Eyak might eventually have been done in by Tlingit.) By the eighteen-seventies, there were only about two hundred Eyak left, mostly in settlements near what is now the town of Cordova, on the Copper River delta. Around 1890, several salmon canneries were built in the neighborhood of Cordova, a development that, in addition to bringing new contagions, disrupted the Eyak's food supply. (A favored fishing technique of the canneries was to use dynamite.)

In 1933, when two anthropologists arrived in Cordova to study the Eyak, the community consisted of just thirty-eight members. The anthropologists, Frederica de Laguna and Kaj Birket-Smith, attempted to chronicle what remained of the Eyak's traditions. Their study, *The Eyak Indians of the Copper River Delta, Alaska,* is at once exhaustive and highly sketchy. "Methods of working stone have been forgotten by the Eyak," the authors note. When they try to ascertain the rituals associated with the onset of menstruation, one Eyak tells them that girls are secluded in a special hut for a month, another says for twelve months, and a third says for six months and not in a hut but in a "special room." Among the many customs that de Laguna and Birket-Smith attribute to the Eyak are building small wooden houses over the graves of their loved ones, observing a taboo against sewing the skins of land and sea animals into the same garment, and burning children's toys in order to secure good weather. Last spring, I phoned de Laguna, who was ninety-seven and living outside Philadelphia. (She spent almost forty years teaching at Bryn Mawr.) I had heard that she had a large collection of photographs from the 1933 expedition, and I asked if she would let me come see them. She said no, she didn't have much time left, and whatever she did have she needed for her own work. "I assure you that growing old is not for sissies," she told me. Six months later, she died.

WHEN MICHAEL KRAUSS began his study of Eyak, in 1963, only six people who knew the language were left. Four of them, including Marie Smith Jones and her sister Sophie Borodkin, lived in Cordova. The remaining two lived in Yakutat, two hundred and fifty miles to the southeast. The six spoke with varying degrees of proficiency, but none used Eyak in daily life. As it happens, the last person who had preferred Eyak to either English or Tlingit had been Smith Jones's mother, Minnie Stevens. When she died, in the spring of 1961, Eyak ceased, practically speaking, to be a living language.

Krauss decided that the best way to preserve what was left of Eyak was to assemble a dictionary. This meant first creating an Eyak orthography. As is the case with many Native American languages, Eyak uses sounds—for example, glottalized consonants and nasalized vowels—that simply are not represented by the Latin alphabet. For most of these sounds, Krauss used either standard linguistic symbols or symbols borrowed from other languages. (A glottalized "t," for instance, is written *t'*.)

Krauss next spent several summers shuttling between Cordova and Yakutat, collecting words. He would point to the parts of his face and ask an Eyak speaker to pronounce the terms for "nose," or *sinik'* (which is really "my nose," since in Eyak you wouldn't say "nose" without indicating to whom, or to what, the nose belonged); "mouth," or *sisa't;* and "eyes," or *silaax.* Then he would ask the same speaker to name the features of his or her face unprompted. In this way, he learned, for instance, that the Eyak have a word—*siniik'adach'uuch'*—for the groove between the upper lip and the septum, for which English speakers have only the clinical term *philtrum.* (The Eyak word translates roughly as "nose crumple.") Occasionally, his informants had rather vivid imaginations and, failing to recall a word, would make one up. When I suggested that the process seemed treacherous, Krauss dismissed this, saying, "Any linguist worth his salt should be able to do the philological workup."

By 1970, Krauss had compiled more than six thousand terms and had come, in his words, "reasonably close to plumbing the depths of living memory in Eyak." The original version of his dictionary ran to thirty-three hundred hand-typed pages. To a nonlinguist, it is nearly incomprehensible:

Demexch' (variant of qemexch'; L certain of authenticity) demexch't (noun, d-class, with -*t*-instrumental suffix) 'soft rotten spot or hole in ice over body or water': demexch'tda'luw 'large treacherous spot in ice' L, demexch'tda'e'q'aGiiyaaq 'don't fall in treacherous spot in ice!' (certain she has heard this).

Linguists typically classify languages according to word order. English is an SVO (subject-verb-object) language: the subject of a sentence usually comes first, followed by the verb and then the object, as in "The woman drove a car." Eyak, by contrast, is an SOV, or subject-object-verb, language, so that a direct translation of the sentence *Lixah daxunh sasheht* would be "The grizzly bear the man killed." As in many Native American languages, both prefixes and suffixes can be added to verbs, resulting in forms that make distinctions not just in tense—did an action happen in the past or in the present?—but also in the degree to which an action has been completed, whether it occurs regularly, what part of the body is affected, and whether it is being done sincerely. Thus, in Eyak, the question "Are you going to keep tickling me in the face in the same spot repeatedly?" is rendered as one word, *xuqu'liitxaaxch'kk'sh.*

It is the differences between English and Eyak that are, in Krauss's view, the reason for preserving it. He asked me if I'd ever taken French. "*Tout à l'heure il sera là,*" he said. "*Tout à l'heure il était là.* Well, what the hell does *tout à l'heure* mean? We have no English word for that. How would you define it—'now plus or minus a short while'? The minute you learn French, you have to learn *tout à l'heure,* and it makes you think differently from the way you ever thought before."

He went on, "Each language is a unique repository of facts and knowledge about the world that we can ill afford to lose, or, at the least, facts and knowledge about some history and people that have their place in the understanding of mankind. Every language is a treasury of human experience. Eyak doesn't give a damn about tenses. But it sure does give a damn about other things, much more than I do. Therefore it broadens your thinking, enriches your ability to understand the world—to deal with reality and experience." I pointed out that it has always been the dream of a certain brand of idealist that mankind would acquire—or perhaps recover—a universal language. Krauss said he had no objection to that, provided that people continued to speak their native languages as well.

"Having the presumption to think we can do with just one language is a mistake," he told me. "Not until we reach the age of enlightenment do we really know what we're doing. And I'd like to see someone claim we've reached the age of enlightenment." I asked Krauss whether he thought that Eyak, like Hebrew, could ever be revived. He said that he rather doubted it. Then what had he hoped to accomplish?

"I've never defined that for myself," he said. "I'm doing it because I'm me, and Eyak is Eyak."

THE TOWN OF CORDOVA, the last home of the Eyak, sits on the eastern shore of Prince Willian Sound, on a narrow bay called Orca Inlet. To the north are the snow-covered peaks of the Chugach Mountains and to the east the marshy plains of the Copper River Delta. Just beyond the delta lies the Bering Glacier, which is the largest glacier in Alaska. Cordova is still a fishing village, and the tempo of life is determined mostly by the comings and goings of various species of salmon. When I went to visit, last August, the silver salmon were running, and the streams were full of enormous, spent-looking fish, lazing around in circles, preparing to die.

The few traces of Eyak culture that remain in Cordova could easily fit inside a suburban garage. A couple of them—a dugout canoe and a spear—are on display at the Cordova Historical Museum. (Also on exhibit are artifacts from Alaska's first oil field, in nearby Katalla, and memorabilia from the town's annual Ice Worm Festival, a Mardi Gras–like affair held the first weekend in February.) Whatever else has survived of Eyak culture is scattered in the woods. One afternoon, I took a hike out to the remnants of an Eyak longhouse. It had been reduced to four logs, arranged in a rectangle. Blueberry bushes were growing out of what had once, apparently, been the floor. Nearby were some graves, marked with Russian Orthodox crosses that were falling into the ground.

The area around Cordova is resource-rich; in addition to the salmon, the oil, and the forests, there is a huge, as yet unmined coal deposit about fifty miles east of town. In recent years, the pressure to exploit these resources has been growing—as has local resistance, much of it led by a group called the Eyak Preservation Council. In an unusual twist, the development battles in Cordova have often pitted one Native group against another. The Alaska Native Claims Settlement Act of 1971 granted billions of dollars' worth of surface and subsurface rights to a complicated network of Native-run boards, and the founder and director of the Eyak Preservation Council, Dune Lankard, who is himself a quarter Eyak, has more than once sued a Native corporation of which he is a member. A few years ago, *Time* named Lankard a "hero for the planet," in recognition of his work to protect the region. Marie Smith Jones, meanwhile, gave him the Eyak name Jamachakih, which translates as "little bird that screams really loud." Although he does not speak any Eyak, Lankard says that he would like one day to be able to file lawsuits in the language.

"We are going to save the language whether anybody likes it or not," he told me. "And we are going to save the land whether anybody likes it or not." He added, "I really think that the Eyak have a lot to offer this screwed-up world, and we are not talking rocket science."

As it happened, my visit to Cordova coincided with the conclusion of a seven-year effort known as the Eyak Language Project. The project was largely the work of a former TV reporter from Anchorage named Laura Bliss Spaan. She first heard about the Eyak in 1992, when she was sent to Cordova to cover the Ice Worm Festival. "When Eyak gets ahold of you, it's really hard to escape," she explained to me. Bliss Spaan arranged for Michael Krauss to give a series of lessons on Eyak grammar, which she videotaped. She then gathered up all the records of the language that she could find—Krauss's hand-typed dictionary, transcriptions he had made of Eyak legends, audio recordings of an Eyak speaker from the nineteen-seventies, video of Marie Smith Jones—and computerized them. Altogether, the archive fit on five DVDs. When I arrived in Cordova, Bliss Spaan was there to deliver the archive to the local cultural council. She offered me an extra set of disks that she had brought along. As I took them from her, I had the odd sensation of holding in my hand all that there was—or ever would be—of Eyak.

The only way out of Cordova is by boat or plane. Flying back to Anchorage and then to New York, I loaded the disks that Bliss Spaan had given me onto my laptop. I listened to Krauss lecture on Eyak vowels, heard Smith Jones talk about how to say hello, and read some

Eyak legends about fantastic creatures. In "Blind Man and Loon," a man's sight is restored by a magical bird. In "Woman and Octopus," a woman falls in love with a spirited cephalopod. There was also a "Lament for Eyak," which Krauss had recorded in 1972. In it, the narrator describes what it is like to be a member of a vanishing people:

> *K'aadih ulah uuch' q'e' iiti'ee.*
> *Sitinhgayuudik siza' iinsdi'aht.*
> *Sitinhgayuu siza' listi'ahtch'aht q'al ahnuu si'ahtgayuu*
>   *q'uh yaan' q'e' distiqahqt,*
> *al iisinh.*
> *Aan,*
> *deelehtdal dlagaxuu,*
> *ts'it dlagaxuu atxstilaht?*
> *Atgaxtalaat.*

> Useless to go back there.
> My uncles too have all died out on me.
> After my uncles all died out my aunts next fell,
> to die.
> Yes,
> why is it I alone,
> just I alone have managed to survive?
> I survive.

Originally published in *The New Yorker,*
June 6, 2005.

Update: Marie Smith Jones died a few years after I spoke to her, in 2008. Michael Krauss died in 2019. Much of what is known about the Eyak language, including a version of Krauss's dictionary, is available online at eyak.org.

ACKNOWLEDGMENTS

All the stories in this book represent collaborative efforts. Literally hundreds of people—scientists, inventors, government officials, and concerned citizens—gave generously of their time and expertise to make them possible. I am particularly indebted to David Gruber and Shane Gero for teaching me how to listen to sperm whales, to David Wagner for teaching me how to look for caterpillars, to David Hackenberg and Dennis van Engelsdorp for teaching me about colony collapse disorder, and to James Russell for sharing with me his deep knowledge of New Zealand's endangered fauna.

Getting to Greenland is always an adventure; thanks to Marco Tedesco, Larry Smith, Tom Wagner, Sune Olander Rasmussen, and Dorthe Dahl-Jensen for making my trip to the Rio Behar and EGRIP possible. Eric Balken was a marvelous guide to the reappearing wonders of Glen Canyon. Søren Hermansen and Roland Stulz were exceptionally gracious in showing me around Samsø and Zurich, as was Hal Wanless when I went to visit him in Miami.

I am grateful to Chuck O'Neal for introducing me to Lake Mary Jane and his campaign to extend rights to central Florida's waterways. Stephen Long patiently explained to me the intricacies of photosynthesis, and Frans Vera equally patiently explained his theory of rewilding. Thanks to Klaus Lackner, Sallie Greenberg, and Noah Deich for explaining to me the intricacies of carbon removal.

James Hansen was a reluctant profile subject; I want to thank him for giving me so much of his time. I am similarly indebted to Christiana Figueres, Amory Lovins, and Sam Wasser for allowing me to intrude on their very busy schedules. I feel fortunate to have met Marie Smith Jones and to have gotten to know Michael Krauss. Thanks, too, to Laura Bliss Spaan, who helped me immensely when I was reporting in Alaska.

It was a pleasure to work with Cecilia Mackay on the photos. Gratitude to Gert Hardeman for his photo of the Oostvaardersplassen.

Almost all the pieces collected in this book first appeared in *The New Yorker*. I am immensely lucky to have worked with many wonderful editors, copy editors, and fact-checkers over the years. Special thanks to David Remnick, Dorothy Wickenden, Henry Finder, Carla Blumenkranz, and the late John Bennet. This book, as a book, would not exist were it not for my editor at Crown, Gillian Blake, and my agent, Kathy Robbins. Their support, encouragement, and guidance have been invaluable.

And, finally, eternal gratitude to my husband, John Kleiner, both for his patience and for his impatience. I would never have finished anything without him.

ABOUT THE AUTHOR

ELIZABETH KOLBERT is the bestselling author of *Field Notes from a Catastrophe, The Sixth Extinction,* for which she won the Pulitzer Prize, and *Under a White Sky,* which was named a top ten book of the year by *The Washington Post.* For her work at *The New Yorker,* where she's a staff writer, she has received two National Magazine Awards and the Blake-Dodd Prize from the American Academy of Arts and Letters. She lives in Williamstown, Massachusetts, with her husband and children.